VIGILAMUS DUM DORMITIS
J. Philip-Sorensen

COMING CLEAN

IAN KIERNAN AND
PHIL JARRATT

MACMILLAN
Pan Macmillan Australia

Proceeds from the sale of this book will benefit the Clean Up Australia Foundation.

First published 1995 in Macmillan by Pan Macmillan Australia Pty Limited
St Martins Tower, 31 Market Street, Sydney

National Library of Australia
cataloguing-in-publication data:

Kiernan, Ian, 1940– .
Coming clean.

ISBN 0 7329 0825 6.

1. Kiernan, Ian, 1940– . 2. Environmentalists — Australia — Biography. I. Jarratt, Phil, 1951– . II. Title.

363.735092

Typeset in 12/14 AGaramond by Post Typesetters
Printed in Australia by McPherson's Printing Group

I dedicate this book to all those people that have supported the establishment of Clean Up Australia and the subsequent Clean Up the World project. They represent the philosophy that if enough of us are prepared to commit to improving all aspects of our environmental behaviour, just maybe we can win the battle to repair the world and live in sustainable harmony with it.

ACKNOWLEDGEMENTS

It has not been an easy task to put this book together and many people have played a variety of roles in its compilation. My parents are described in some detail in the book and I would like to pay tribute to my Mum, Leslie Kiernan, who, apart from bearing us and raising me and sister Ann through the difficulties of World War 2, has always been a wonderful model of enthusiasm and optimism. The fact that at eighty-three years of age she is still pursuing an energetic and successful painting career is reflective of her determination and joy of life.

My daughters Sal and Pip I thank for the true friendship that we share. Thanks too for putting up with a father who has always been preoccupied with the next adventure, challenge or disaster, and who probably wasn't around that much when he was needed.

To my wife Judy, loving, stoic and steady as a rock, who has always been so patient and supportive of me and of my goals. The day we agreed to marry I said, 'but please no restrictions on my sailing'. It was only fifteen months later that I came home with the news that I had a sponsor for the BOC Solo Around the World Race. With hardly a blink she stepped in to support me in every way possible.

I do not have that many true friends, but I have always treasured those friendships and I regard them as longlasting. The fact that some of these friends have played some part in this book with comments about me is interesting to say the least! Their observations are here as spoken, even though in some instances I was tempted to play the role of unofficial editor. It seems I'll have to get square with them in some other way. Perhaps that will mean yet more visits to the bakery!

The directors of Clean Up Australia and Clean Up the World were drawn from friends of many and varied talents, and I would like to thank them and the dedicated and skilful team that carry out the day-to-day duties of our organisation for doing so much to make the ambitious goals that we have set so far a reality.

Especially, I would like to thank Kim McKay for her friendship, and for all her help in making the BOC race a reality for me. It was her vision – in fact, brilliance – that in due course largely led to the implementation and success of my simple ideas of Clean Up Sydney Harbour, and to the subsequent formulation of Clean Up Australia. Then, not content to rest there, coming up with the idea to take these successful formulas to the world. Clean Up the World! A modest program I am sure you'll agree. Talented, hard-working and tenacious, Kim has played a major part in making this book a reality.

Finally to my co-author Phil Jarratt (or 'JARAY' as we know him) – another person of huge talent that falls into the category (and enjoys the dubious honour) of having me as one of his mates. I have to admit that casting back into the past to provide him with the details of my life to date has not been easy. It was often painful to remind myself of bits of that life that I had chosen to forget. It is in some ways interesting that so much of the transcript for this book was recorded with Phil on board *Maris* sailing back from Hobart after the 1994 Sydney to Hobart Race which is where this book starts and finishes.

There is so much that I want to do, so many more challenges to take up, so much money to raise to allow us to progress and finally succeed with what we have started with Clean Up. I will leave it there for now so that I can get on with living Chapter 19.

IAN KIERNAN

This book would not have been possible without the support and enthusiasm of our friends at Pan Macmillan; most particularly I would like to thank Deborah Callaghan and Amanda Hemmings, and for his sensitive editing, Roger Milliss.

Special thanks to my friend Paul Wright, who took the jacket photo, and to Rob Mundle and Bruce Stannard, whose specialist writing about yachting and yachtsmen was an inspiration as well as a source. Finally, the spirit of two people I never had the privilege of meeting pervades almost every page of this book. Through Ian, the Bilge Rat and Jack Earl became very real to me, and as I tried to work out what makes Ian Kiernan tick, I felt their guidance.

PHIL JARRATT

CONTENTS

FOREWORD

'What can I do about it?' 'What can one person do?' Behold the opt out, cop out, pass-the-buck questions that most of us use to excuse inertia. From time to time, however, someone comes along who demonstrates, loud and clear, that the I'm-a-drop-in-the-ocean argument simply doesn't hold water, if you'll forgive the mixing of two aqueous metaphors.

And I've been privileged to know and work with two of them.

The first. A young lawyer who lived in a boarding house in Washington. He had no funding, no back-up staff and communicated via a pay phone in the hall. Yet he terrified the most powerful corporations in America, forcing them to change both their behaviour and their products, and proceeded to have a comparable impact around the world. Not only has the lawyer, no longer young, saved countless lives but he's also protected millions from injury.

Not content to be David to just one Goliath, he went on to tackle other giants just as fearlessly and effectively. There was a time when he was the most famous and admired person in the US, when he would have been an irresistible candidate for the presidency. Yet he chose to stay in his Washington boarding house, using the pay phone, showing people how to be critical consumers. Arguably he was the inventor of single-issue politics, laying the basis for, for example, the environmental movement. He certainly created the worldwide phenomenon of consumer protection as he forced industry after industry to consider its responsibilities to consumers and the community.

His name was Ralph Nader and he did all this, and more, without raising his voice. Ralph used a combination of legal training and commonsense to confront recalcitrant management, to force them to take responsibility for decisions that were hurting people. It all began when he wrote a book called *Unsafe at Any Speed,* accusing Detroit of knowingly marketing cars with dangerous defects. There was an immediate and immense reaction to Nader's book and to his personal style. With his shambling gait and rumpled suit (I once accused him of rumpling his suits on purpose) he seemed to have wandered in to Washington like a character from a Frank Capra movie, a latterday Mr Deeds. There he spoke simply and quietly about corporate sins of omission and commission, confronting Detroit with hard evidence of cars that fell apart, of petrol tanks that exploded on impact, of designers who refused to consider safety when creating new models.

Thanks to Nader the world now wears seatbelts. Thanks to Nader, cars are now designed to crumple progressively in accidents, so that the impact is absorbed and, as far as possible, the capsule that carries the driver and passengers is left intact.

Ian Kiernan is Australia's Ralph Nader. Like Ralph, he's something of a loner. Whilst Ian's the most gregarious and popular of people, his career as a solo yachtsman reminds you that, at heart, he'd prefer to be away from the crowds. He's never happier than when he climbs into the car with his faithful cattle dog and heads off somewhere. Anywhere. Like Nader, he's been utterly unchanged by his success and by the honours and accolades and tributes that have poured in on him. Like Nader, he's operated on the smell of an oily rag, losing rather than making money from his efforts. Like Nader, Ian inspired people to take issues into their own hands, not to wait for someone else to solve their problems. Like Nader, Kiernan will have nothing to do with buckpassing or making excuses. When he sees a problem he simply rolls up his sleeves and gets stuck into it.

Like Nader, Ian's not into histrionics or the politics of abuse. His technique is quiet, reasonable and rational. He mounts arguments in preference to tossing grenades. And, like Ralph, he's transformed a local campaign into a national campaign and, finally, into a considerable global achievement.

Like Nader, Kiernan works within the system. He prefers negotiation to confrontation, consensus to conflict. Like Nader he is determined to decentralise, encouraging people to get themselves organised at the local level. Instead of seeking to gather power into his hands, Ian wants people everywhere to have power for themselves and he shows them, by example, how to get it.

Ralph and Ian are decent, modest and finally ordinary human

beings. 'Ordinary' isn't a put-down. It's not used pejoratively. What I mean is that neither Ralph nor Ian was born rich or well-connected and neither has a 200-plus IQ. What makes them effective is that they are recognisably decent, straight-forward people with none of the trademark traits of the professional celebrity. People instinctively trust Nader and Kiernan, safe in the knowledge that they're not pushing a secret agenda involving the achievement of personal power.

Ian Kiernan was upset by the muck he saw in the water when he was sailing around the world. He decided to do something about it and rang a few friends and organised a campaign to clean up the beaches around Sydney Harbour. The rest is history, culminating in a campaign that is now cleaning up over 100 countries. Millions upon millions of people are involved. Values and attitudes are changed.

Nader started with dangerous cars. Kiernan started with dirty beaches. In a sense, the specific issue was irrelevant. Nader might have chosen any other industry or category of product to demonstrate corporate carelessness and worse. And Ian might have chosen another social problem as his catalyst.

For Ian, garbage is a device, an excuse, a focus. What the Clean Up Australia process does is to gently politicise individuals, groups and communities, making them realise that individual responsibility matures into collective responsibility and, finally, community spirit.

I've met and interviewed a lot of self-proclaimed Messiahs. (You'd be amazed at the number of people who've contacted me, insisting that they were Jesus Christ and it was my job to promote their second coming.) In truth there are some saviours around, some model Messiahs whose claims for themselves are less hubristic and whose contributions are incomparably greater.

In announcing Ian Kiernan as Australian of the Year, I described him as 'the greatest garbo since Greta'. Ladies and gentlemen, I'd like you to meet someone I'm proud to count as a friend. That little Aussie battler, the quiet achieving Ian Kiernan.

PHILLIP ADAMS

INTRODUCTION

On Boxing Day 1994, Ian Kiernan — Australian of the Year, Captain Yukky Poo, solo circumnavigator of the world, well-known larrikin and good bloke — sailed out of Sydney Heads into a light sou'easter and a sea of confusion.

Some 370 boats were jockeying for position around the marks in this, the fiftieth anniversary running of the Sydney to Hobart, one of the world's greatest offshore yacht races. But the huge number of yachts stealing wind from each other's sails and churning the Harbour to a frothing mass of white water was not the sole reason for Kiernan's state of mind. As he steered his beloved yawl *Maris* to the mark, his mates and veteran yachties Tom Stephenson and Dick Hammond calling the shots and the sail changes, the Australian of the Year, aged fifty-four, was nearing the end of the busiest year of a busy life. It is difficult to overestimate the pressures placed on Kiernan by a professional schedule covering such a broad sweep of activities, a sometimes chaotic personal life and a fundamental inability to say no to any worthwhile cause. In the days leading up to the race start he had on his plate a daughter's wedding, chairmanship of the Sydney committee of the BOC Challenge round-world yacht race (a relatively minor role cascaded into controversy through the dramatic rescue of competitor Isabelle Autissier), preparation for the launch of the 1995 Clean Up Australia campaign and a swag of speaking engagements taken on to help cover the cost of the Hobart race. Oh, and preparation of *Maris.*

It was here that Kiernan fell down just a little. 'The bloody radio didn't work,' said Tom Stephenson, Kiernan's sailing mate of more than twenty years.

'He needed a project manager,' said crewman Marty Farrand. 'There were so many things left undone.'

Kiernan's sailing friends have a version of Chaos Theory to explain his apparent disdain for the normal conventions of competitive yachting. They believe he thrives on the challenge of the calamitous circumstance, and actually longs for those moments when a cool head, a piece of string or wire and a lot of guts can save the day. But even as they talk in frustration about him, his sailing mates can't help but reveal a grudging respect and a love and admiration for this garrulous, egotistical, totally crazy, cheerfully obscene, extraordinarily energetic, compassionate and caring man. They know that if chaos rules his life on land, it is the sea that sorts him out. Always has.

In 1974, when he sailed *Maris* out of Sydney Heads in similarly confused circumstances, amid rumours around the pubs that her keel was solid gold, he was simply Ian Kiernan. 'Bicky' to his mates, 'the trendy developer' to his mortal enemies in the Builders Labourers' Federation, 'that slippery bastard' to the banks and finance companies, running scared on the precipice of the Sydney property collapse.

At sunset on a September evening Kiernan eased *Maris* out of Mosman Bay while his great mate and mentor Jack Earl, the sailor and maritime artist, blew a farewell on a Darnley Island conch shell. For Kiernan it was more than a farewell; it was a turning point in a life which to that time had embraced capitalism and hedonism in about equal measure. He would never completely change his stripes, but from that day on his life, his marriage, his relationships with people, his view of the world would never be the same.

Kiernan is a creature of his time, and a creature of Sydney, a port town which grew up along the same confused but exciting lines between the 1960s and the '90s. When he first started to turn a penny on real estate, Sydney's central business district was encased by a horseshoe of slums, a kind of cultural moat between the places where men worked and the safe, homogenised places they went home to. For a young man of vision, Sydney was a place filled with excitement and opportunity. There were barriers to be broken down, prejudices to be overcome, millions to be made.

For Ian Kiernan, Sydney was also the town that straddled the most beautiful harbour in the world, a place of quiet, clean beaches, rocky points lapped by water teeming with table fish, and broad, wind-whipped sailing runs with the salt spray in your face and a cry of joy in your heart.

The war babies came of age in a time of full employment, allegiance to the Queen, fear of Asian neighbours and European food, a strong work ethic and the expectation of holding one job for an entire working life. Kiernan wanted to make money, do everything, go everywhere and have fun all the while. He was thus at odds with the establishment from the start, despite being a part of it. If the first part of his adult life was spent proving to the business community that conventional wisdom was not necessarily the best, the second part (from the age of thirty-five to the present) has been spent proving to himself that the creation of wealth is not the measure of human worth, and to others that individuals, not governments, change society; that one person can make a difference.

Writing this book has not been easy. Sometimes one gets the impression that Ian Kiernan couldn't organise a piss-up in a brewery, to use his own vernacular. And yet this is the man who had a vision of motivating people to clean up their own mess in Sydney Harbour, and turned it into an international network of thirty million people in ninety-five countries working together to solve environmental problems.

Time is his greatest enemy, organisation its ally. Kiernan is habitually late, chronically distracted. His normal attention span is two to three minutes, except on those subjects about which he is passionate. And these change day to day, week to week. A mediocre student, he has become obsessive about understanding the things that will influence the future of the planet. Despite his disorganisation and his eccentricities, he has a practical, analytical mind and he has been able to distil the information he gathers from environmentalists, scientists and academics, and put it to work for his cause. Getting him to put it to work for this book has been the hard part.

I met Ian at Sydney's Cruising Yacht Club in 1986 at the launch of radio station 2MMM's *Spirit of Sydney* campaign, the cornerstone of radio mogul Rod Muir's multi-million dollar assault on world ocean yacht racing. An audacious lad from Tasmania who travelled at a million miles an hour, dropping a bank note with every step, Muir had dreams of owning a fleet of the world's fastest yachts, simultaneously campaigning for the coveted titles of the world's most prestigious races. The 60-foot welded aluminium *Spirit*, purpose-designed by the Ben Lexcen team which had brought Australia the America's Cup just three years earlier, was scripted to win the 1986–87 BOC Challenge, a gruelling single-handed race around the great southern capes, beginning and ending in Newport, Rhode Island, the scene of Alan Bond's and John Bertrand's great triumph.

Spirit's sailor was to be Ian Kiernan, a builder/developer who had lost his fortune in the '70s and had clawed back, winning dual reputations as one of Australia's best restoration builders and most experienced solo yachtsmen. I had known Rod Muir peripherally for a few years, and while I admired his boundless energy, I could barely understand a word he said. Kiernan was a different kettle of fish, a friendly, outgoing, laconic raconteur and bon vivant that one warmed to immediately. We became friends as I documented his preparation for the BOC Challenge, and I also became friendly with Kim McKay, a dynamic young public relations consultant who had helped Kiernan in his search for sponsorship and was now working on the Challenge.

In August 1987, a few months after Kiernan had finished the race in sixth position, McKay wrote to me, saying that Ian had begun to write his life story. She enclosed a sample chapter. 'Unfortunately, he is no Ernest Hemingway,' she noted. No, but I liked his lusty account of life as a slum landlord in the middle of the Sydney vice wars, and the contrast with his solitary and courageous life as a solo sailor on a storm-tossed sea. Kiernan even had a name for the book — *Twice Around The Shark Pool*, a reference to his circumnavigation of the world and his earlier run-in with the sharks of the financial world.

The book might have come about in 1988, had not another project become all-consuming. As he sailed the final leg towards the BOC finish line in Newport, Kiernan had been appalled at the amount of rubbish that littered the surface of the once-pristine ocean. Alone on the sea, his long-held thoughts about the growing pollution of his beloved Sydney Harbour began to crystallise. Why not try to do something about it?

The extraordinary results of these solitary thoughts over the past six years have created another volume in the saga of Ian Kiernan's life, and another reason his story had to be written.

This is not a ghost-written autobiography. It is a collaborative work based on more than sixty hours of taping of the subject and those who have been involved in his life, plus many days which the two of us have spent poring over dusty scrapbooks, photo albums and ships' logs. Like any man of action, he has his critics, and where their view is at substantial variance with his own — or else where other people have thrown additional light on what he says about himself — I have interpolated their comments in the text, because I think it is important for a man like Kiernan to be seen in an all-round fashion, rather than through what might be perceived as a personally sanitised version of his life.

My own feeling about Ian Kiernan — and it is hard to be objective about a friend — is that he is a great Australian, not just because he has achieved so much and taught so many to take responsibility for their own environment, but because he typifies a breed of person who may be rare but will never be extinct. Proud, noble, generous and brave, hideously flawed and unmistakably Australian.

So pull your bar stool closer, fill up your glass. Bicky is holding forth.

Phil Jarratt

CHAPTER ONE

WAR BABIES

I was born in Sydney a year after the war began in Europe and a year before it began in the Pacific. Neither declaration had much of an immediate effect on sleepy Sydney — the second certainly had none on sleepy Ian Bruce Kiernan — but the war was to change the course of my parents' lives, to make them both resilient and fiercely independent of spirit, and perhaps some of that rubbed off on me. Or could it have been our stubborn Scots blood?

My father was born George Arthur Carrick Kiernan in Melbourne in 1903, the son of Scottish immigrants. His father died suddenly when George was very young, and his mother's mother came out from Scotland to help the young widow take care of the boy and his sister. Granny apparently had some money but Kiernan family lore has it that she soon fell prey to the unscrupulous bloodsucking merchants of Melbourne and frittered away her fortune on silly investment schemes. Like much of the Kiernan family history, it's a good yarn, if a little short on evidence.

As soon as he was old enough, George left school and supported the struggling family through the years of the Great War with a succession of odd jobs. When his mother had remarried and the family circumstances improved, he became more selective about his work, eventually joining the retail firm of G.J. Coles & Coy as a trainee manager.

Coles Emporium in Melbourne was the brainchild of George

and Keith Coles, who had returned from the war and seen a great opportunity for retailers in a decade of hope and optimism. Theirs was the first age of consumerism, and the company grew quickly, forcing the Coles' to seek bright-eyed and ambitious young employees who would grow with it. George fitted the bill perfectly. He was well-groomed, and as well-dressed as his meagre savings would allow, he was quick to learn and he had a pleasant manner.

By 1934 George Kiernan had survived the Great Depression and had become one of the key managers at Coles, and at the age of thirty he was sent to London to oversee the company's buying operations. Life in the City of London suited George admirably. He made friends easily, enjoyed the clubbish camaraderie in which business was conducted (usually over cigars and port) and spent many happy weekends down in the country fishing for trout or playing golf with business acquaintances. On one such weekend he enjoyed the company of Coles' London accountant at his home in Buckinghamshire. After dinner, when the two men were playing billiards in the study, they were joined by the accountant's daughter and her friend, an attractive lassie down from Scotland by the name of Leslie Katherine Mary Bruce.

I'd be delighted to say it was love at first sight across a crowded pool room, but my mother remembers thinking only that George had a kind face and moved delicately for a big fellow. Nevertheless she was chuffed when her friend Audrey phoned her at the London Polytechnic where she studied art, and reported: 'My dear, I've something killingly funny to tell you. George Kiernan wants to meet you again.'

Mother was twenty-four years old when George began to court her. She was born in Glasgow where her father, Norman Bruce, worked as an engineer, but both her mother's and father's families were from Edinburgh. In the '20s Norman Bruce accepted a position with the London and Rugby Engineering Company and the family moved first to a flat in the city and later to a house in Buckinghamshire. Norman worked in the sciences but his heart was in letters. He read extensively and spent long evening hours in his study with a fountain pen composing erudite 'Letters From London' for publication in his old school magazine. The focus on arts and literature in the Bruce home may well have contributed to Leslie's decision to study art at the Polytechnic. Certainly, she showed a natural aptitude for painting, something she has gone back to with much joy and enthusiasm in her later years.

Once he had made up his mind about Leslie Bruce, George Kiernan didn't muck around. He drove down to Buckinghamshire, played a round of golf with Norman and asked for his daughter's hand in marriage over a gin at the nineteenth hole. The next day he popped the question to Leslie, adding that of course they would be sailing for Australia as soon as possible after the wedding. My mother had another plan. 'It's not fair on Mummy and Pop,' she said. 'Let's marry sooner and spend a few months here before we leave to get them used to the idea.'

They were married in December 1936, honeymooned in Switzerland, then took a flat in South Kensington. For six months Leslie continued at the Polytechnic while George wound up his affairs in the city and made arrangements for his successor. At weekends they would drive down to Buckinghamshire for golf and Sunday lunch with the Bruces. But in the summer they had to sail. Hitler was already on the march in Europe and thinking Britons were becoming concerned about what the future held. Norman Bruce may thus well have felt some relief that a good man was taking his daughter to the far side of the planet, to a place where they at least spoke English and revered the House of Windsor.

Leslie's first sight of Australia was the dock at Fremantle, from which point she imagined Melbourne must only be a short hop. (Geography is still not her strongest subject.) The Kiernans sailed on around the Bight and finally into Port Phillip Bay. Her introduction to Melbourne was a gruelling round of dinners and cocktail parties, during which she met the entire Coles family and most of their employees, but just as she was beginning to feel her way around the city, George came home one evening and announced he had been transferred to Sydney.

George elected to drive up the long, narrow Hume Highway, deviating from it to show Leslie the city of Canberra, the new capital of Australia. She recalls: 'It was such a tiny wee city, right out in the middle of the bush, but I was quite fascinated by it. We drove up to the Parliament House right in front of the beautiful lake, and I remember thinking how new and modern everything was.'

They arrived in Sydney after dark and checked into the Hotel Australia. The next morning George said: 'Right, I'm off to work. You go and find us a flat to live in.' Mother found an estate agent who drove her through the Eastern Suburbs where she caught glimpses of a glistening body of water. When the agent showed her a ground-floor flat she was quite adamant: 'Oh no, I must have a view of the lake.'

They took a flat in Point Piper and Leslie quickly learned the topography of the city. The terminology took a bit longer. My sister Ann was born in 1938 and the family moved to a bigger flat. George was becoming convinced that war in Europe was inevitable. He wrote of his concerns to his father-in-law, and Norman responded, agreeing. However, he wrote, 'if war comes you must not send Leslie and the child back here while you go off to fight. No one in England will be safe.'

That year George joined the Militia, accepting the rank of first lieutenant with the 1st Division at Victoria Barracks. When war did come, in September 1939, he enlisted in the AIF immediately, despite the fact that he was almost thirty-six years old. Through much of 1940 he moved from training camp to training camp, returning home to Leslie and Ann whenever he could, particularly when they discovered Mother was pregnant again.

With his departure for the battle zones imminent, George decided that the family had to have a house of its own. He told Leslie: 'If I'm away I don't want you and the babies thrown out of a place at a moment's notice. You must have a home.' They found one, a rambling, jerry-built 1920s structure called 'Gillingwood' in Fitzwilliam Road, Vaucluse. George would have bought it that first spring morning they looked at it, but there was something more pressing. Mum had gone into labour.

I was born on 4 October 1940 in St Margaret's Hospital, Macleay Street, Potts Point, and when mother and child were ready we were taken home to Gillingwood, which was the only permanent home I ever knew as a child.

In early 1941 George was transferred to the 8th Division with the rank of captain, and in April, when I was just five months old, he sailed for Malaya on board the *Neiu Holland*, part of the biggest troop ship convoy to leave Australia. The *Queen Mary*, *Queen Elizabeth* and *Mauritania* all sailed on to the Middle East, but the *Neiu Holland* stayed in Singapore for some weeks, before heading up to Malacca, then further up the Strait of Malacca to Taiping, where George and his men alighted. After the Japanese entered the war in December 1941, the unit joined up with the Argyle and Sutherland Highlanders at the Thai border to search out infiltrators, but were forced back by the advancing Japanese to Ipoh, Johore Bahru and finally Singapore. During this retreat a bomb blast killed George's sergeant as he stood next to him. George was hurled into a ditch and broke his leg, but fortunately it had time to heal before the cat-and-mouse game with the Japanese quickened its pace.

The division regrouped and George's unit left for Burma by way of Java, just a day ahead of the fall of Singapore. But in Java they were ordered to stay and prepare to defend against the impending invasion. During this critical period my father was promoted in the field to major, but it was only a matter of time before the Japanese invaders forced a surrender. First the Dutch forces capitulated, then the British and Australians were taken prisoner. George was sent by ship back to Singapore, then on to Changi and finally to the dreaded Burma Railway, where he was to spend the greater part of his three-and-a-half years as a POW.

George had been away for more than fifteen months before

Leslie received word from the Army that he was 'missing, believed POW'. She threw herself into the task of caring for two young children and tried not to let her imagination work too hard. Finally, another message from the Army: 'Confirmed POW'. Her relief was immense, but who knew what the prisoners were having to endure?

After long months of agony, my mother received a postcard from Burma, but there was no scenic view or 'Wish you were here'. On one side was her name and address, on the other a bald printed message in which a set of options had been ticked. 'I have been in hospital.' 'I have not been in hospital.' 'I am well.' 'I am not well.' It was signed, as my mother still recalls with great disdain, like a cheque — G. Kiernan. No 'Lots of love', no 'Miss you badly'. It wasn't much at all, but at least it was something.

During the three-and-a-half years George was a POW, Leslie received just three of these cold, clinical cards, and that was the only evidence she had that he was alive. Although she had no family to turn to, other than in letters, she had a strong network of neighbours, and a couple of the women shared her predicament. 'Don't worry about the words,' they encouraged her. 'Look at the strength of the signature. If his hand is not feeble, he's all right.'

Leslie wrote letter after letter to her husband, and eventually a package of them was sent back undelivered. Undeterred, she had us children photographed in Sunday best and sent that off to George too. In one of those small miracles that occur in a war, or perhaps in one's memory of one, one of my father's men happened to be in the administrative office of the camp when a string-bound parcel of letters was delivered and dropped on the desk. Seeing a familiar name on the top letter, the soldier grabbed a handful of mail from the top of the pile and stuffed it down his baggy shorts. In this way the photograph of my sister and me made its way to the '30-kilo camp' on the Burma Railway in one of the darkest hours of the Pacific War.

George found wood and perspex (off the windshield of a plane) and a piece of wire, and he fashioned a little frame. He kept it with him for the rest of the war, and my mother has it still, although the wire hinge has rusted away and a little rotter who was visiting cracked the wood with a deft blow.

The loneliness my mother must have experienced over those war years is something she always shrugs off lightly, but I can only begin to imagine how she must have felt — a young woman with two children cut off from her family and the world she knew, while the man that she loved was another world away, at the mercy of an enemy proving with every news despatch to be entirely lacking in that particular quality. There were Kiernans in Melbourne, of course, but she barely knew them, and getting to Melbourne in wartime was a feat in itself. Instead she relied on her Vaucluse neighbours, who were marvellous, and spent a lot of time walking a stroller along the foreshore.

I was too young to remember much of this, but apparently we all slept in one room for comfort. Perhaps my earliest memory is of waking in the night in this room to find my mother reading by the light of a soft lamp. 'Go to sleep,' she said. 'Everything is fine.'

Everything was not fine. The Japanese had bombed Darwin and the people who lived along the foreshores of Sydney Harbour feared they would be cannon fodder in a Japanese invasion. These fears were found to be justified on the night of 31 May 1942, when Japanese midget submarines entered the Harbour. Those who could were taking their children to the country. K.F. Coles had taken an avuncular interest in us in my father's absence and arranged for Leslie to take a flat in the town of Orange, across the Dividing Range. She took us up by train and we evidently made our home there for almost a year. I don't remember a damn thing about it, but I'm sure we were totally safe, Orange being fairly low on anyone's invasion priority list.

What I do remember in the war years is the experience of real country life at a property called Dockairn, near Bathurst. Dockairn was the country estate of the Moran family, who had four rough and tough boys who knew how to load a slingshot and catch a rabbit. John Moran was the head of Colgate-Palmolive and a business friend of my father's. He had seen action in the First War and knew what Leslie was going through. Thus we were frequent weekend and holiday guests. I loved Dockairn with its gorgeous old homestead and gardens with ornate Georgian fountains. We'd sit on the verandah in squatter's chairs and watch the blue wrens come in to drink, and sometimes the great dust storms of the central west sweeping across the plains.

There weren't many males around when I was a little fellow, so John Moran was important to me, as was Dr John Bell-Allan, a family friend. Dr John was always there to patch me up when I came a gutser, but I think he was also there to patch Mum up too, with a kind word and a shoulder to cry on when the going was too tough to bear.

Eventually the war swung in our direction. In the near north the Japanese were on the run and beginning to panic. They broke up the POW camps into more manageable groups and George was sent to an officers' camp across the border in Thailand (then Siam). Then in August 1945 the Americans dropped atomic bombs on the Japanese cities of Hiroshima and Nagasaki, thereby unleashing the most extraordinary example in history of what man will do to his fellow man and to his own environment. The war was over in a week, and my father was on his roundabout way home. The celebrations around Australia were uninhibited by the sobering statistic that of some 22,000 Australians taken prisoner by the Japanese, 7,964 did not come home.

Siam having been liberated, George went to a 'fattening-up' base in Bangkok, where he stayed with White Russian friends he had made before the war while working in the cushy job of embarkation officer. The first thing he did in Bangkok was to find a tailor of repute and order a cream linen suit for the climate. He might have been without money or muscle, but he was not about to look tatty.

In the summer George came home. John Moran came to the house to take Mum to the wharf while Ann and I stayed at Gillingwood for the more intimate family reunion. The troop ship

Circassia pulled in with my father waving from the deck along with so many other gaunt and frail men. When she finally held him in her embrace, my mother thought, 'My God, there's not much of him!' George's opening gambit was perhaps not his finest. 'Where did you get that hat?' he enquired.

As John Moran's car came down the street, my excitement got the better of me and I had to rush to the toilet, thereby missing the melodramatic moment. When I came out he was sitting in the living room with a cigarette and a mug of beer. I suppose we hugged and kissed, but all I remember is the strangest feeling, an excitement I couldn't explain. It was like Christmas — he opened a suitcase and produced toys from the Orient for both of us — and yet it was not. My Dad was home, but I had no clear idea of what a dad was.

I think it took Leslie a while to get to know him again too. She had got in an evening cook so that they could enjoy that first night together, and ordered roast beef and Yorkshire pud. She was flabbergasted when George, who had loved the stodgy tucker of the Mother Country, played with it and left it. It took weeks to get him eating well again (brains were apparently his new favourite) and there were other changes too. His legs and arms were deeply scarred from tropical ulcers, his insides were torn about with hookworm and dysentery, and the bomb blast which had killed his sergeant had left him completely deaf in one ear.

Over that long summer I came to know my father, somewhat hesitantly at first. He spent a lot of time sitting in the garden and smoking. I was too young to understand what he'd been through, but I knew instinctively when to leave him alone with his thoughts, and as the weeks went by and his strength grew, he became more active and outgoing. He was a military man in bearing and inclination, as well as in the experiences of the previous six years, and he liked a regimented life. Dawn was for swimming, and he'd shake me from my sleep and take me with him to Nielsen Park, where he would swim slowly and deliberately along the shore, watching me in the shallows with one eye as he took in breath.

Not satisfied with my rate of progress, Dad took me along to Mr Seabrook's swimming lessons at Bondi Baths, then later on to Alf Vockler at Watson's Bay. Alf's tools of trade were a kicking board and a canvas harness on a length of rope. He'd walk along the edge of the pool holding the rope while I kicked and dog-paddled for all I was worth. The chafing inflicted by this harness was truly horrific. I'd walk home very gingerly after each lesson, but my dog-paddling improved and Dad cut a deal with me. Once I had learnt to swim overarm to the extent that two lengths of Nielsen Park were not a problem, he would take me to the surf.

My father took delight in these small steps for mankind but great leaps for a young Kiernan, because in some ways he was going through the same steps himself, growing in physical strength each day and re-establishing himself as the head of the family. I wish I'd been a little older that summer, old enough to understand his strength of character as he built himself back into the man he'd been before fate dealt him a pretty lousy hand. It wasn't as bad as some, of course, and as if by way of consolation, the poker schools of Sydney treated him fairly well in the ensuing years.

At the end of the summer George went back to work at Coles, where he took over the buying department, not that there was much to be bought. Still, he'd been an officer in the war and a manager of long standing before that, he'd served in both callings with distinction and he was now treated with respect, and rewarded accordingly. We were not rich but by postwar standards we were comfortable. Although he had eaten little of the cook's offerings on his first night home, George now insisted on a permanent cook. 'Dammit,' he said to my mother, 'When I come home in the evenings I don't want you in the kitchen, I want you having a drink and a chat with me.'

So we got a cook. And another cook, and another. Some were good and some were atrocious. Lily, a Hungarian Jew who had fled her homeland, was my mother's favourite. 'Who taught you how to cook, Lily?' my mother asked after a particularly delicious meal. Lily said: 'Our cook's at home.'

Although he mixed with the captains of industry and knew most of the decision-makers of the day, George Kiernan was neither a snob nor an elitist. He had the common, earthy touch which had endeared him to his men and fellow POWs (many of whom remained friends until their deaths) and allowed him to walk easily in all quarters of the city. He liked a drink, a smoke, a bet, a fish, a game of golf, a funny story. He liked to dine at the best restaurants in Sydney, but he also liked to murder a steak or a chop on the barbie at home. No doubt people said George Kiernan was a good bloke, and by the end of 1946, while I didn't fully understand him, he had entered my pantheon of heroes.

That was the summer my grandparents came out from England, bringing with them the kilt that my Uncle Alistair had worn as a lad. As a mark of respect for one of the heroes of Dunkirk, I was forced to parade in this un-Australian get-up while the Box Brownie captured irrefutable evidence for the Bruce clan over the seas. Despite this unfortunate episode, it was a happy summer of picnics and boat outings on the Harbour. Mum hadn't seen her parents for ten years, and for half of that time, with the world at war, it seemed they might never meet again. There was a bit of the dour Scot in Norman Bruce, but he rarely had a smile off his face that holiday.

Nineteen-forty-six was also the year I made a somewhat inauspicious start to a singularly inglorious school career, entering a kindergarten at the end of our street. Before the school year began I had commandeered a box tree which shaded a bus stop in front of our house. High in the branches I kept several Milo tins filled with water, which I used to drench neighbouring kids the moment before they boarded the bus. They would cast their eyes skyward, see only leaves fluttering in the breeze, curse and mutter and climb aboard. It was a game I thoroughly enjoyed until I bombed a fierce-looking woman in quasi-military attire who wheeled around and saw me, then knocked door-to-door until she had established the parentage of 'that little fair boy'.

As if the punishment I received at home was not enough, I

arrived at my kindergarten to discover that the harridan in the greatcoat was none other than Mrs Gomm, the head of the school.

I started at the prep school at Scots in 1947 when Norman Bruce was still in town. He was quite interested in my suitability to games, particularly cricket, and was appalled at the mediocrity he saw in the nets. My mother has a clear recollection of him watching a practice session at the college in which balls passed through legs and bats were flailed wildly. Norman Bruce tut-tutted and declared, 'Oh Lord, this will never do.'

Scots College, like many schools in those postwar years, was going through a period of readjustment. The teaching staff had been severely depleted during the war years, which meant in some respects that it was the blind leading the blind. Certainly that was the impression I had in my brief time at the school. The masters and I were not entirely compatible.

There was also the possibility that sickness was the cause of my poor scholastic performance. This was something investigated thoroughly by my mother, who was concerned at one stage that I might have had polio. I didn't, and no one ever diagnosed a serious ailment, but in my time at Scots I suffered constantly from colds, sore throats, upset tummies — you name it, I had it. My father began to talk about sending me off to the country. Our friend John Bell-Allan's son had been school captain at The Armidale School and the doctor spoke highly of its teaching staff, its facilities and its rural outlook.

Late in 1950 we drove up to New England and had a look around. It looked pretty good to me and my parents were terribly impressed with the rather grand red pebbled driveway leading to a collection of brick buildings with terracotta details, designed in 1894 by Sir John Sulman. Arrangements were made and when school resumed for 1951 my parents took me to Central Station at dawn and packed me onto the school train, filled with rowdy boys in grey serge, some of whom were to become my friends for life.

CHAPTER TWO

LEARNING THE ROPES

I sat still for barely five minutes at a time for the entire duration of the long steam train journey to Armidale. For a ten-year-old it was incredibly exciting. Although there were many old hands on board, the camaraderie of the school train (not unlike the troop ship) was based on a shared fate, and although new chums like me didn't have a clue what that fate might be, we were granted temporary equality while we discovered each other's names, strengths, weaknesses and sporting heroes.

The further we travelled from Sydney, the more excited I became. The stations rolled by . . . Werris Creek, Tamworth, Walcha Road, Uralla . . . I had grown to love the bush during our wartime sojourns to Dockairn, and with every passing hour I could feel myself being drawn back to its comforting vastness. Perhaps that's a little strange coming from someone who has spent most of his adult life in the city, but I suspect my affinity with the bush comes out of the same part of the brain as my affinity with the ocean. In both places you can feel totally, wonderfully alone.

We were met at the station and taken by bus to the school at the other edge of town. Armidale then wasn't the city it is now, but it was the seat of the northern squattocracy and the houses and homesteads had the style of the gentry about them. The Armidale

School itself was quite grand to look at, not so when you examined it for creature comforts. There were about six main buildings on the property with a chapel at the gate and playing fields front and back. Inside, the rooms were austere and draughty, which was useful in summer when the temperature could climb into what we knew as the 90s, but bloody awful in winter when the New England frosts often didn't melt until afternoon when the cold westerlies would blow across the plateau. There was very little in the way of heating.

There were about 250 boys in the school and all but half a dozen were term boarders. We lived in dorms of about forty hard beds and a shower and toilet block. When the morning bell sounded the drill was shower, PT, breakfast, chapel, then straight into class. I didn't mind the PT (physical training) nor the breakfast, and I could even put up with chapel, but that morning shower was a terrible thing, at least until I had been at TAS long enough to work my way up the pecking order. Although 250 boys had to shower over a period of about twenty minutes, the hot water from the boiler-house lasted only for three or four minutes. After that it was briefly lukewarm and for the final ten minutes it was icy cold. It was thus important to be in the top half of the social order and desirable to be in the top twenty per cent. I got there eventually, but not until I'd steeled myself for a thousand or more icy showers.

TAS was undeniably disciplinarian and authoritarian, but in the main it was the kind of discipline that at least made some sense. You knew the gamble you took when you broke the rules and the masters were hard but fair. There were exceptions. I loathed one master, who had been a professional wrestler before he became a teacher, or so it was rumoured. At the age of fifty-four, I would like to think that I could look back on my years at TAS and think of something nice to say about him, but I can't. He was a deadset bloody mongrel. He taught maths and geography and we had 'Black Fridays' when we had him for three periods before recess and another two after lunch. Although he never brought his wrestling ability to bear in the class room, he was fond of the cane to the point of kinkiness, and when he caned you, he left white marks across the palm that lasted all day.

Mind you, that master wasn't alone in his passion for the cane. Another master liked to draw a chalk stripe across the seat of the pants and try to raise dust with each stroke.

When we weren't being caned nor suffering under cold showers, we had fun at TAS. In addition to the academic subjects, we learnt bush craft and there was a heavy emphasis on team sports. Most of the boarders had bicycles and on occasion we'd be allowed out to ride into town. At weekends we were granted 'bush leave', which meant we could ride unsupervised as far as fifteen or sixteen kilometres out of town to make a day camp by the rivers or creeks. Sometimes at night we'd sneak out of the dorm and ride off on orchard raids, filling our pillow slips with apples and cherries. At one point these raids reached epidemic proportions and the farmers jacked up. They lay in wait for a group of boys (fortunately not my group) and beat a sense of rights of property into them.

Our other anti-social behaviour involved hurling huge round river stones at each other, or firing marbles out of bunger guns. There was no malice involved in this potentially life-threatening activity. It was simply sport. The older fellows got quite carried away with this, and moved from bunger guns to serious munitions, making their own gunpowder in the science lab. The war games ceased with a major explosion, when one poor guy blew part of his hand off. The list of injuries also included sight impairment, partial deafness, hundreds and hundreds of stitches and a lot of wounded pride.

I arrived at TAS a year after all that fuss with the explosives. I enrolled for first year of high school, whereas Ian had started in the junior school. He was an absolute arsehole, a real smartie. His attitude was that he was a smart boy from Sydney and fellows like me were country lads, away from home for the first time, and to be despised. It takes the first term to get these

> *things sorted out, and after a few bouts of fisticuffs we became very good mates. We'd go on bush leave at weekends together and get up to all sorts of pranks. On one occasion we borrowed a couple of rifles from the armoury and shot at birds . . . anything that moved, really. We should have been expelled.*
>
> Graham Hartigan, school mate.

I made some fantastic mates at TAS, and as I got a bit older I took to spending holidays with them on their family properties rather than going home to Sydney. I learnt to ride horses and handle stock and fell in love with the lifestyle, so much so that I switched courses to include agriculture for a while, thinking that I'd go on the land. 'That's all very well,' my father said, 'But we haven't got a property, as you may have noticed. You'll be starting at the bottom.' I stuck with agriculture, but my yearning for a life on the land eventually gave way to thoughts of a city career.

> *Ian used to come out and stay on our family property. We'd go shooting and drinking and carrying on. He loved the land, loved animals, loved to shoot them. He wasn't much of an environmentalist. He liked shooting things and throwing his rubbish around everywhere. By the same token, he did have a love of trees and a love of the rural life. We just had different priorities in those days, I suppose.*
>
> Graham Hartigan

In my secondary school years I discovered girls and rugby, and became equally passionate about both. At Armidale in the 1950s it was pretty easy to get a game of rugby, not so a pretty girl. Nevertheless, we rarely stopped talking about them as we smoked Rothmans and handed around a bottle of beer in some secret hideaway. The New England Girls School (NEGS) was our sister school and we had combined social evenings and dances, but the girls were watched like hawks. You had to make whispered clandestine arrangements to meet at weekends, or at night if you were particularly daring. As puberty took hold of my better judgment, I found myself tiptoeing out of the dorm after lights out and riding across town to dangerous and achingly exciting rendezvous, from which not even the chill of the New England night could deter me.

My third-year maths teacher, Ken Jago, got me into rugby. The headmaster was the coach of the First XV and Mr Jago had the Seconds. His tactic was ridicule, and while I can see now that it was a pretty risky ploy, it had the desired effect on me. Every time I dropped a pass he'd call me 'Cacky Kiernan', which infuriated me. I resolved to show the bastard and never drop a pass again. I was not a gifted player, but Ken Jago drew the best out in me and with a lot of determination, in time I made the First XV. At TAS this carried with it the exalted status of school hero and the awesome responsibility of 'setting an example'. This meant a lot to me because I'd been a bit of a rebel and, through rugby, I felt I'd become a better person, more responsible, more caring, a team player.

Much of the credit for the moulding of the new Kiernan went to my housemaster, Jim Graham, who was something of a new-wave teacher. Jim preferred talking to caning, encouragement to ridicule. When I was on my way up to the First XV he recruited me to help him train the Under-13s, which not only gave me more time on the football field, but forced me to think about my ball skills in order to pass them on.

> *Jim Graham and Ian were very matey. Ian used to wash his car for him at weekends. I remember one Sunday after chapel Ian had the keys to the car and we drove it about fifty kilometres out into the bush and back. Neither of us was old enough to have a licence, of course. Jim Graham must have known, but he never said anything.*
>
> *As for football, Ian wasn't much good at it really. He didn't have a natural talent for ball games at all. But he worked at it, he was a real trier.*
>
> Graham Hartigan

Given my stature as a member of the Firsts and my strong belief that I had pulled myself up by the boot straps, I was somewhat shocked to find in my final year at TAS that my name failed to appear on the list of senior prefects, prefects or even monitors. It seemed like a calculated insult after the hard work I'd put in on character building and personal growth. Maybe I was still more wicked than I thought. I tried to put this out of my mind, particularly after I was appointed a monitor to fill a mid-year vacancy, and to get on with my lessons and football.

Although we were a high-spirited lot, there is only one incident I can recall that blotted the copy book of the First XV. We played just one round each year, touring to the GPS schools in Sydney or playing home games against them. After one big match the TAS Old Boys, which is still a very strong outfit, hosted a dinner for us. Since we'd notched up a notable victory, there were many toasts and a few of us overstepped the mark in the consumption of alcohol. Naturally the headmaster found out about the ruckus and we were summoned to his office the moment we returned to Armidale. There we were threatened with expulsion, and worse, the cancellation of our next game! Fortunately, given our position on the points table, sanity prevailed and our contritions were accepted.

I sat for the Leaving Certificate in 1958, just after my eighteenth birthday, and scraped through with the 'gentleman's pass' — four Bs. This was not exactly a triumph, and on my father's insistence I sat for posts in the new year and eventually matriculated.

During my high school years I'd learnt to sail, and it became an enduring love despite the long months I spent away from the beach and the Harbour. I first sailed in the school holidays the summer I turned thirteen, but initially the sport had to compete with surfing for my affection. I'd been a keen surfer from the moment Dad conceded I could swim two lengths of Nielsen Park and took me over the hill to the real surf at Bondi. Soon I was diving off the rocks at Bronte to save the swim out, and body-surfing to the shore. Later I bought one of the first balsawood malibu surfboards, brought to Australia by a team of Californian lifeguards here to compete in a special surf carnival held in conjunction with the 1956 Melbourne Olympics.

I enjoyed surfing — still do — but as I became more proficient at sailing, I could feel myself being swept along in a strong current leading further and further from the shore. I crewed in Saturday races at the Watson's Bay Sailing Club and later joined the Vaucluse Yacht Club. With my young sailing mates Tony Holt (now on the board of Clean Up Australia) and Peter Hemery (now chief executive of the Royal Sydney Yacht Squadron), I would watch the Tasman Seabirds on the Harbour and yearn to be a famous ocean racer, like Boy Messenger or Trig Halvorsen.

By the time I was fifteen I had to have a boat, and I worked at chores all through the summer to make the money to pay for it. When I was back at school in Armidale my father phoned to tell me there was a Veejay for sale. He told me the price and, knowing I hadn't saved enough, allowed me to wallow in disappointment for all of thirty seconds before he said he'd make up the difference. Now I was a serious sailor.

Dad was very good to me that year. I don't know whether it was the fact that he'd passed fifty, or that he'd finally quit Coles to go into business for himself, but he seemed to want to do things with me more than he ever had before. We travelled to New Zealand together, just two guys on the lam, and he taught me the tricks of travel as well as using the opportunity to impart what he regarded as the more important elements of his philosophy of life.

The following year I sailed on my first real yacht, Peter Luke's *Wayfarer*, which had been a competitor in the very first Sydney to Hobart race in 1945. Peter and his father Monte Luke were Sydney's most famous portrait photographers of the time. Thanks to the two of them, a great many Sydneysiders of my vintage have attics full of highly embarrassing sepia-toned photographs of themselves dressed as Little Lord Fauntleroy. Peter was a respected sailor, and on hearing of my love of sailing, arranged for me to crew on *Wayfarer*, an unforgettable experience and my introduction to the activity that has shaped my life.

By the time I finished school in 1958 I'd moved up to sailing Bluebird class yachts and was starting to improve, but sailing wasn't my only interest that summer. There were the posts for one thing. University wasn't on my agenda, but Dad was insistent I should matriculate just in case I changed my mind later. But he had his own plans for me too.

In 1951 Dad had left G.J. Coles after more than twenty-five years of service and with a host of international buying contacts. With the patronage of the Coles family, he set up G.A.C. Kiernan & Co Pty Ltd, with offices in Castlereagh Street and contracts to import salmon from Russia, socks from Britain, tools from Germany and chinaware from Japan. He dearly wanted me to join him in the business, but not until I had received the appropriate training. Where else but at G.J.Coles? So I started work in the glassware department at Store No 130, just along from my father's office in Castlereagh Street.

The job may have been boring but the rest of my life was not. I was sailing in the summer, playing sub-district rugby with the Colleagues in the winter and chasing girls full-time. I had a Volkswagen at first, then a Holden. I wore shorts around the waterfront and moleskin trousers and a tweed jacket to rugby functions. It was all very conservative but it seemed to work with the women. My mates and I cut a swathe through the pubs and clubs of the Eastern Suburbs — the Golden Sheaf and the Royal Oak at Double Bay, the Astra at Bondi, the Latin Quarter in Pitt Street or the all-night Greek clubs in Redfern and Newtown. I drove like a maniac and collected speeding tickets at an alarming rate.

When I wasn't off with my mates or chasing girls, I'd sometimes accompany my father to the 33 Club, an illegal casino in Darlinghurst. Most nights this place would be packed with Sydney's leading lights from the fields of politics, business, sport and crime. Some regulars at the 33 Club fitted into several of these categories at the same time. The then-NSW Premier Sir

Robert Askin, for example, was a politician, businessman, sportsman and criminal as well as a few other things. Dad and I would nod good evening to them and get on with our illicit pleasure.

Things started to go bad at Coles when they moved me from glassware to groceries and then to gardening supplies. The thought of selling potted petunias to little old ladies every day filled me with despair. I'd already virtually given up on my evening accounting classes and spent as much time as possible with my workmate Gordon Marr, wishing we were somewhere else. My finest moment at Coles came when a section manager told me to stick some sale banners on the main display window. I thought he meant inside the window, and attempted to thread my way through a maze of very attractive store dummies fitted out in underwear and negligees. As I reached to put the banner up I knocked a dummy and set off a domino effect. I examined the wreckage and prepared to face the wrath of the section manager, but then I noticed something far worse. There was a crowd of people outside the window laughing at me.

I fled the window and soon after fled Coles.

My father agreed that I'd perhaps had enough experience of retailing and set me to work in the storeroom at G.A.C. Kiernan, but I'd barely acclimatised myself when he announced that my sister Ann and I were to accompany him on a voyage around the world. Although the big credit squeeze of 1960–1 was about to hit, the importing business had up to this point been kind to George, and he felt he deserved a lengthy break from it. I was perhaps less deserving, but no less enthusiastic. So while my mother flew off to spend time with her family in Scotland, my sister, my father and I sailed on the *Willem Rhuys* (later the *Flotta Lauro*) to New Zealand, then via Pitcairn Island across the Pacific to the Panama Canal and up to Miami. Here we left the ship and hired cars and caught trains cross country, eventually making our way to New York.

We then sailed from New York to Naples, a voyage made memorable for me by Dianne Collins, a beautiful young New

Yorker of Italian extraction. Ours was the classic shipboard romance, except that I couldn't shake her out of my system when we parted in Naples. Dad, Ann and I travelled on to Rome, where we watched the 1960 Olympic Games, then to London where some job training had been lined up for me with Lloyds and a dockside customs clearance agent. The work was interesting enough, but I couldn't wait to finish my commitments and hop on a plane back to New York, where Dianne and I were reunited.

That was a magical time but it wasn't doing much for my career, whatever direction that might take next. I kissed Dianne goodbye and flew home to Sydney.

Although G.A.C. Kiernan Pty Ltd was chiefly involved in importing, my father had set up a subsidiary company to enter the export market. While importing quotas under licence provided a safe and consistent income, it was also low-profit and, to my mind, fairly boring. There was a fairly lucrative trade in the export licences themselves at that time, but George Kiernan had a very strong view about the morality of that. Instead he put his risk money into a joint venture with an old desert-rat opal miner named Frank Brown.

Frank was about seventy and suffered from sandy blight, the desert equivalent of snow blindness. The condition made him squint so badly it distorted his leathery features. In the city he always wore the same outfit, a threadbare suit, check shirt and tie and an Akubra pulled down over his eyes. Out on the opal fields he would remove the suit and don overalls, but he always wore the tie.

My father's accountant, Ernie Hill, introduced him to old Frank at a time when there was a developing market for opals in Germany, Japan and the US. Frank knew opals and Dad had the contacts, so Brown Kiernan was formed and on my return from the arms of Dianne Collins I was seconded to work with the new company, learning from Frank how to grade and value opal. The work was interesting, Frank was a beaut old bloke and I looked forward to our field trips.

We'd drive down through Hay and Renmark to Port Augusta,

where Frank kept a battered '48 Chevy ute for the rough ride to Coober Pedy along the Woomera Rocket Range road. This track was covered in gibbers as big as twelve or fifteen centimetres in diameter, and at night the desert was alive with big kangaroos. It was impossible to miss them all, and every trip we'd add a few dents to the Chevy, but the greatest danger was that a roo would slide up the bonnet and through the windscreen, where it would go beserk.

Coober Pedy in those days was hardly a town at all. There were just a few corrugated iron and timber buildings visible on the scrubby hill. There was no pub because the miners well knew the danger of a constant source of beer in the arid desert climate. Instead they would bring in a truckload of beer every Friday afternoon and drink their way through it by Sunday night. Monday it was back to work no matter what shape you were in.

Frank had a dug-out on Central Hill. After the First World War, diggers had been attracted to the opal fields by the prospect of quick money. Having learnt about underground living in the trenches of Europe, they looked to dugouts as cheap and heat-resistant dwellings. Frank's dugout was a two-cave affair with separate shafts to disperse the heat from our tilley lights, stove and kerosene refrigerator. The bunks were hard and the furniture crude, but we were cool and comfortable enough in there. Certainly our dugout was no match for some of the miners who had struck it lucky and built huge underground palaces with air-conditioning and deep freezers and flash cars parked on the gibbers above.

There was no open-cut mining at that stage and the mines consisted of very basic vertical shafts dug by hand to the seam of opal at either twenty-five or fifty-five metres below the surface. These shafts had a manual or petrol-driven winch over the top of them and the miners rode down to the seam in a bucket. From there the various lateral shafts were dug, using a rather devil-may-care bush technique. The miners would drill the face, plug it with a charge of gelignite, light the fuses and sprint like crazy to the bucket and be winched to the top before the charges blew. When the dust cleared they would return to the shaft, remove the fill by hand and hopefully expose a bright-coloured seam of valuable opal.

The mines radiated out from Coober Pedy in bands known as the Three, Five, Seven, Nine and Eleven Mile Pegs. These produced opal of varying quality and one of the traps for young players was an opal from the Eleven Mile that, despite its fine appearance, would crack and eventually break up with climatic variations and was virtually worthless. Frank knew his rocks and never fell for that one.

Often when we arrived at the fields Frank would be carrying as much as £40,000 cash in his money belt, so that we could inspect and grade the various parcels offered to us and make a good deal for payment on the spot. Of course it became known that we had large sums of cash on us and we always carried guns for protection. Frank never made a big thing out of this and our guns were not displayed, but we kept to ourselves most of the time, particularly during the weekend booze sessions, and I think even the toughest of the miners had a respect for Frank and knew that he was not to be taken lightly.

I enjoyed the work and the frontier atmosphere of Coober Pedy, but the only relief from our fairly spartan existence there came in the form of the occasional arrival of young adventurers. These prototype backpackers were often European and often female. As a civilised chap from the city, I felt it behoved me to offer the safe haven of our dugout to the prettiest of the visitors. Frank would grin and mutter and shake his head, and then go back to squinting at his opal specimens.

By 1961 G.A.C. Kiernan had developed considerable two-way trade with Japan, importing china and exporting opal. It was decided that I should accompany Dad's major buyer, Jim Dalrymple, on a business trip to meet our agents and suppliers. We flew to Bangkok, then Hong Kong and finally Tokyo where Jim, an experienced Asia hand who knew the ropes, came into his own as a guide to the social and cultural life of Japan. We worked hard by day but at night we ran even harder, hitting the bars and nightclubs and having a wonderful time. I lost count of how many glasses of coloured water we bought for the beautiful hostesses who were employed to sit at the tables of Westerners and be sociable.

At one club I became friendly with a girl named Kyoko, beautiful and quite tall by Japanese standards. One night after she had finished work we ended up at a traditional Japanese hotel. I was not aware of the complex Japanese bathing ritual in which you wash and scrub and bucket off before you enter the formal bath tub. I will never forget the look on Kyoko's face when she walked in and found me lathering up amidst a sea of soap bubbles in the huge tub. Fortunately the culture shock passed and it wasn't long before she was in the tub and giggling with me.

My job at G.A.C. Kiernan had its moments, but the credit squeeze had affected us, as it had most businesses, and my father started to scale down operations. He was approaching sixty, wanted to work less and enjoy life more, and more importantly, he wanted to snug the company down so that it could ride out the economic downturn and be around to prosper under my stewardship later on. This was all sensible and admirable, but it didn't fit in with my plans. With every new skill I learnt in business, I became more anxious to push ahead, take risks, make bigger profits. And I felt I was in the wrong business. The construction industry was beginning to transform Sydney and I wanted to be a part of it. In 1962 I moved out of home and quit G.A.C. Kiernan.

The move came first. My grandmother followed Norman Bruce to the grave and I was left a small inheritance with which I made a deposit on a tiny, run-down apartment in Darlinghurst Road, Kings Cross. I put a solid door on and fitted a heavy lock, Mum and my sister made the curtains and I moved in, subletting to a couple of mates. Mum reckons I still took my washing home for her until Dad put his foot down, saying: 'You've left home now, so off you go.'

But he harboured no grudge about my leaving the family business, recognising that my future lay elsewhere. I applied for a cadetship with Concrete Industries Monier and was one of three successful applicants for their 1963 intake.

Monier's then-general manager, Jack Davenport, was a man of vision who had evolved a cadetship system through which young people were given job experience across the whole range of the company's operations, while studying management and other related subjects at technical college at night. After a qualifying period we were then given major projects to manage under the scrutiny of Jack's divisional managers.

My first job was in the yard at Monier's Villawood factory on the western fringe of Sydney. It was hard physical labour but I thrived on it, building up to a physical peak for my footballing career with Colleagues. Our team had turned into a formidable outfit and we would sometimes tour for country or interstate matches. On one tour I can remember incurring the wrath of the pilot as we charged up and down the aisle of a chartered DC3, but he later got into the spirit of things, buzzing the homesteads of our country hosts before we headed home.

After my stint in the yard I moved to Monier Pipe, making concrete pipes in huge steel moulds. The moulds were two metres long and two metres in diameter, and we'd use an overhead crane to lift them onto a set of four trunnion wheels, then turn on the power and watch the mould rumble and roll on the trunnions. A dump truck would drop a couple of tonnes of wet concrete beside us and we would begin shovelling it into the mould, using a length of reinforcing rod to settle the concrete in the mould and a trowel on a stick to smooth the surface of the pipe. We would make two or three of these in a day, but sometimes the speed at which we had to work created problems. I still bear the scar from one accident when I slipped on the wet floor and hit my chin on the steel mould. The factory manager, Jack Pell, looked at the wound, pulled it together and taped it with Elastoplast, patted me on the shoulder and sent me back to make another pipe.

When I'd completed my basic training at Monier I was made job captain on the State Office Block, which was being built in

Macquarie Street, Sydney. Monier's contract was for the manufacture and supply of the structural concrete cladding for the twenty-storey building. My job involved programming the production of the panels, overseeing their manufacture and liaising with the on-site building team. It was my first real management job and I found it immensely rewarding, but the project had exposed a lot of raw nerves. With two major government jobs on at the same time — the State Office Block and the Sydney Opera House — the unions had a great deal of leverage and regularly bunged on stoppages. The American builder, Perini, who had introduced some new and quite experimental building techniques, was losing a fortune on the construction and applying the pressure down the line. The State Office Block was not the only thing called 'SOB' during that period.

As an introduction to construction management, I couldn't have done any better than the State Office Block. In my usual brash fashion, in early 1965 I felt that I had nothing more to learn as the job neared completion and handed in my notice at Monier to take on the position of assistant manager with a company called Davleco, which made roadbuilding and construction equipment. If my career to this point had been punctuated by the brash moves of an impetuous youth, this one represented the exclamation mark of a slightly foolish one. There were great opportunities for me at Monier, but they weren't happening quickly enough. Davleco offered immediate advancement and I grabbed the position without knowing much about the company nor the people who ran it.

The manager was a slick salesman and a great networker who had built the business out of his contact list. His best mate was the maintenance manager who felt he should have been given my job. I was rooted from the start, and it didn't take long for a situation to arise where a small mistake was enough to hang me. I ordered some pumps and was a bit sloppy about the detail. The supplier complained to the maintenance manager who complained to his mate, the boss, and I was gone. This was the first and only time in my life I have been sacked, and I was extremely annoyed about the way it was done. One of the directors of the company flew up from Melbourne and gave me two weeks' notice. I was shaken by this, but I've always felt that notice is bullshit, that if you've gotta go, you've gotta go now.

I walked out of the Davleco office that afternoon and never went back.

CHAPTER THREE

THE PROPERTY GAME

Fortunately I got the sack in summer and was able to retreat to the waterfront to lick my wounds. I had become sailing-crazy and this was a great opportunity to make the most of a 28-foot Herreshoff ketch I'd been given the use of.

The ketch was owned by a family friend who rarely got to use it. The deal was that I would keep it in good nick for him and take it out as I pleased. What an opportunity this presented! Once I had mastered the boat on the Harbour I started making small sojourns out into the Tasman, usually just day trips up to Pittwater or down to Botany Bay, but enough to give me a taste for ocean sailing.

Often I would take my girlfriend of the time, whose father, an experienced yachtsman, never tired of warning me that the sea was a dangerous and unpredictable place, and that I should show it respect. I paid him as much attention as most young men do their girlfriends' fathers, but on one Pittwater voyage he was proven right. We were enjoying a spinnaker run home when a southerly buster blew up with such intensity that I had to turn back and run before it to seek shelter back up the coast. When the initial intensity of the buster abated we were able to tack home, but we made a pact never to tell anyone how close we had been to coming unstuck.

Although I had a free boat and a couple of tenants paying the mortgage on my flat, the cost of my active social life soon meant that I had to look for work. I was pretty handy by now and picked up piecemeal jobs easily enough. The McDowell's department store building in the city was being air-conditioned, so I started working with the plumbers and riggers on the installation. It was a strange job. The equipment was ancient, the installation was Mickey Mouse and the contractor seemed disinclined to pay us. A few of us had to go along and sort the guy out.

Around this time I met an amazing rough diamond of a bloke named Bill Bridges. Bill was a Damon Runyon-style character, a former jockey who had gone into real estate and was making a killing buying up terrace houses around the edge of the city, doing a cosmetic renovation and selling them at a good profit. From the moment I met him at the Royal Oak in Double Bay, Bill made me laugh. He'd made and lost a couple of fortunes and enjoyed several wives, and he was a great storyteller. He made the real estate game sound like fun as well as money for jam. 'Mate, mate . . . you pick 'em up for nothing, bloody nothing, you give 'em a whitewash, burn the furniture in the backyard and sell 'em to the next bugger.'

He offered me a job and I accepted on the spot.

> *I was the first bloke into Paddington. Painted it white and made a fortune. I'd started to move into other parts of the inner city when I met Ian. I offered him a job doing up our places. He was a great and tenacious worker, let me tell you. I took a shine to the feller. He was ambitious all right. He was always quizzing me, always looking for a tip. He said, 'If you were me, where would you start?' I told him I'd go into Darlinghurst with my ears back, and buy everything I could find.*
>
> Bill Bridges, real estate salesman

In the early 1960s the ring of 'slums' around the city of Sydney was still a 'no man's land', the province of the vice lords, the inner city poor and the migrant families who had come out to a country of promise and found they could only afford to buy homes where Australians feared to tread. The endless streets of terrace houses had been built in the second half of the nineteenth century. Many of the more elaborate ones, dripping with cast-iron lacework on their street facades and with an unused fireplace in every room, were the work of Victorian-era speculators, while the plainer rows of identical houses adjoining corner block factories or warehouses had been built by employers to provide rental accommodation for their workforce.

After the First World War, however, Sydney's urban spread began in earnest, with new suburbs eventually sprawling over the plains to the foot of the Great Dividing Range. By 1960 the developed area of Sydney extended over 1000 square kilometres, while more than a quarter of the population (mostly in the new suburbs) had no mains sewerage in their homes. The Great Australian Dream involved a red brick bungalow and a lawn big enough for a Hills Hoist and a game of cricket. The cramped terraces of the inner suburbs were for society's underbelly.

Between 1946 and 1949, the Chifley Labor government instituted a series of radical social welfare reforms designed so that all Australians could see 'the light on the hill'. Similar policies were followed by Labor governments at State level. One of these reforms in New South Wales was rent-fixing or rent control to protect the poor from unscrupulous landlords. Although this was well-intentioned, the net effect was often disastrous. Three and four-bedroom terraces had their rents fixed at as little as fifteen or twenty shillings (one or two dollars) a week, which by the 1960s meant that the owners couldn't sell them and couldn't afford to maintain them. In some of the larger terraces 'head tenants' had nice little subletting rackets going. The head tenant, who may have not even lived on the premises but kept a room spare to satisfy the Landlord and Tenant Act, would pay the fixed rent, then collect about ten times that amount from his sublets.

Things were beginning to change in the slums, and Bill Bridges had been astute enough to reap the benefit. He would buy houses for as little as $1,500 (£750 in the old currency) and send me in with a team to 'do a mocker' on them. Mostly this consisted of a coat of paint and the replacement of anything broken beyond repair. Bill had a formula for making a profit out of these purchases and part of it was that the renovation budget was virtually non-existent. The other part was buying at the right price, usually at auction and often from the estates of landlords who had been bled dry by rent control.

Although they may not have realised it at the time, Bill and other dealers like him were starting to change the face of inner Sydney. As the writer Craig McGregor noted in his 1966 book *Profile of Australia*: 'Many of (the terraces) are sub-standard, narrow-fronted residences which have long been crumbling into disrepair and are due to be torn down in slum clearance schemes or to make way for expressways, but others have suddenly become desirable again because they are so close to the city centre and have been restored and repainted by artists, writers, New Australians, speculators, graziers seeking town houses and those members of the arty set who like to be fashionable even in choosing where to live.'

When I had worked for him long enough to see the profits he was making, Bill Bridges convinced me I should sell my flat in Kings Cross and buy one of his specials. 'Mate, mate, I'll look after you with something good.' He did. I picked up a modest profit on my flat and bought a dilapidated rent-controlled mansion in Little West Street, Darlinghurst for $6,000. It was a large, freestanding, two-storey Victorian building with verandahs top and bottom and cast iron all around. You walked in the front door and down a central hallway with cedar doors leading off to either side. Each of these led to two double rooms which had been divided into two tenancies. There was another wing out the back, huge yards and a rear lane access. It had once been quite grand, but like so many rent-controlled premises, it had been frozen in time at about 1927. The place had twenty-three tenants who didn't live in squalor but came pretty close to it. They shared bathrooms, and most of the tenancies, taking advantage of the slack fire ordinances, had gas rings for cooking. There were white ants, leaks and rising damp everywhere you looked.

The head tenant paid me £9 a week and sublet the mansion for a total of £60. Obviously the name of the game was to get him out and collect that rent myself, but this was no easy process. I moved into the room the head tenant kept for himself, commenced legal action to get him out and started to fix the place up.

At about the same time Bill Bridges and I went our separate ways. I had become a liability because I couldn't help myself once I started working on Bill's houses. I started taking out walls and putting new balconies on, which I saw as value-adding. Bill saw it as unnecessary expense. In the end he said, 'Look mate, I can't afford to do this. If it's what you want to do, why don't you go out on your own?' It was a good tip. Bill has copped a lot of flak over the years, but I always found him as straight as a die, and we parted good friends.

> *I called him a 'change-of-life restorator'. My philosophy was always leave something for the wife to do, a coat of white paint was all you needed really. So we had a bit of a difference of opinion, but we stayed mates.*
>
> Bill Bridges

In 1965 I had started dating Judy Allen, a tall, bright, gregarious blonde girl with a wicked wit. She had been on the fringes of my social scene for some time, and in fact I'd taken out her flat-mate a few times. When the flat-mate went overseas I asked Judy out. She was educated at the Presbyterian Ladies' College (PLC), highly motivated and, just a year older than me, already a senior account executive at the advertising firm Brown and Bruce. She was an impressive woman, never more so than the night a few months after we started dating when she presented me with a blue cattle dog pup called Alf. This was the start of two great love affairs, and the one with Alf was to last the longer.

Judy and I were married towards the end of 1966 at Scots Chapel in Bellevue Hill. My young offsider in the building game wanted to see the wedding so I lent him my truck and he drove down in his blue singlet with Alf in the cabin beside him just as Judy and her father arrived at the church. She caught a glimpse of my truck and my dog cruising past and started to feel confusion building within her. Fortunately, when her father led her into the church there I was at the other end of the aisle.

The wedding reception was held at the old Belvedere Hotel in Darlinghurst and we stayed there the night, just me and Judy and Alf at the foot of our bed. In the morning I ordered lamb chops for breakfast and hid Alf in the bathroom when it was delivered. No doubt the room-service waiter was a little perplexed when he returned to collect the plates to find them licked clean and the bones crunched to pieces.

Judy and I lived in a rented flat in Paddington while I conducted my legal battle to get the head tenant out, and continued to establish myself as a contract builder, first as a sole trader, then as IBK Constructions. There was plenty of building work for someone with initiative and I concentrated on the high-profit business of building factories and warehouses in the rapidly-expanding light industrial areas of Sydney's west. I'd set up my team on a factory site, then race back across town to oversee other small teams busy renovating terraces on the city fringe. Although this wasn't as lucrative as the factory work, it gave IBK Constructions exposure across the entire building spectrum. I was very big on signs. My Volkswagen truck was heavily signwritten and at every site I'd put up an IBK Constructions sign. Soon I had one of the best-known small building companies in town, and the work started to roll in.

Eventually I was able to reach an out-of-court settlement with the head tenant, which gave me the green light to improve the place in Little West Street. The first thing I did was to build a little flat out the back and Judy and I moved in. Living in a street where the neighbours picked through our garbage was perhaps not quite what a nice girl from PLC was used to, but Judy had even more ambition than me, and she could see the end plan.

I started negotiating with the twenty-two remaining tenants and got several of them to sign leases at market rent, while others eventually died or moved out when I purchased their rent control tenancies. As a Labor voter all my life, I felt quite strongly about my social responsibilities to these people and tried to treat them with respect and fairness. There was one rather difficult old tenant named Slim. He played hardball, but I respected him for that and we became friends. When he got cancer I was one of his few regular visitors in hospital over his last months.

But there were some cases where I wasn't prepared to be tolerant. A couple of years later, after I'd moved out, the old lady upstairs was the last rent control tenant at Little West Street and I got word she had died. In those days you could actually will a rent-controlled tenancy, but the assumption was that whoever inherited it was in similar economic circumstances and was planning to live there. By the time I heard about the death the relatives had changed the locks and staked their claim to the tenancy. I decided to fight.

Early next morning I climbed onto the roof and removed a sheet of iron. I cut a hole in the ceiling and dropped into the empty premises and changed the locks again. Next day the relatives came back, borrowed a ladder and changed the locks yet again. I was really mad now. Next day I had 2,000 bricks delivered and my bricklayer Siggie and I bricked up every door and window, from the inside. When the relatives returned they were relieved to find the locks unchanged . . . until they opened the door and ran into a brick wall!

Judy became pregnant at the end of our first year at Little West Street and started dropping hints that we should move somewhere more suitable for bringing up a child. I couldn't see how. I was working seven days a week to make the money to renovate the old mansion, but it wasn't in good enough shape to sell at a profit. Judy had been talking to Harold Baker, a solicitor and rugby mate of mine. Harold explained that we could revalue Little West Street, based on the improvements I'd made, and take out two mortgages on a new place. It seems so simple now, but in 1967 real estate gearing was an unknown science to people like me. A little light went on in my head and a new door opened.

We bought a bungalow in Mosman, overlooking beautiful Balmoral Beach, for $20,000, and our first child, Sally, was born there on 11 February 1968. The birth of my daughter had a profound effect on me in ways that are perhaps perfectly natural. While the female does the nesting, the male squares up to his responsibilities as provider. I was already working very hard, but I redoubled my efforts in order to provide for this additional responsibility. And I became extremely aggressive in my business dealings.

A week after the birth I borrowed some money from my father and some from Judy's father, and bought four terrace houses and a vacant factory in Riley Street, East Sydney for around $40,000. I applied Bill Bridges' formula to the renovation and Harold Baker's to the financing. Because I had my own building team, the renovations were quick and effective and in smart time I had turned a row of derelict houses into a rentable proposition. Brian Syron, the Aboriginal actor, was my first tenant, and within a month rents were bringing me an income well in excess of the mortgage payments and I was able to pay back the family debts. I was on the way to building a portfolio of city property.

I quite liked my role as terrace house landlord. On Saturday mornings I'd do the rounds with Alf, collecting the rents, having a chat with the tenants and fixing any problems they might have. This was a great thing to do because it told me so much about what people were looking for in an inner-city house. Without really trying, my builders and I had stumbled upon a style of housing that had real appeal to the new city dwellers. While younger people had started to see the advantages of city living, in

1968 they still expected to live in a hovel and sacrifice creature comforts for proximity to the office. We did our places with a bit of style — sandstock brick, window boxes and other little touches — and spent money on the bathroom and kitchen. I had no real budget and no use for architects or interior designers. I had a bit of an eye and I'd do my own sketches and give them to a drafting service. Not everyone who saw the 'To Let' ads liked what they saw, but we had no problem filling them at better than market rent.

To keep renovation costs to a minimum I used to keep an eagle eye on the auction columns. One day I noticed a washing machine going through FR Strange, where I happened to know the auctioneer. On the day of the auction, however, I was busy on a concrete pour and turned up just as bidding started on Lot 37, which I was relieved to have knocked down to me for $70, a terrific bargain. Then bidding began for Lot 38, a Westinghouse washing machine! Lot 37 turned out to be a six-inch pipe threader, an unusual piece of equipment but certainly very useful if you were building an oil rig. Highly pissed off with this turn of events, I bought the washer for a good price anyway, and left the auction wondering what I was going to do with my pipe threader. The next morning I got an answer in the form of a phone call from a rather embarrassed chap who had arrived after the threader had been knocked down and had then tracked me to my office. He said he had a specific use for the threader, new ones cost a fortune and second-hands were as rare as hen's teeth. He knew there'd be a premium, but would I consider an offer? I couldn't believe my luck. He had no qualms at all about parting with $200 and I had a free washing machine and money in the bank. Talk about getting out of jail!

In the mid-1960s the East Sydney vice wars erupted again. The area had been the scene of gang conflict since the 1920s, when illicit booze and cocaine fuelled the tempers of the so-called Razor Gangs. In the '60s many of the terrace houses were controlled by rival vice bosses and as many as 100 brothels competed for business. One tactic used by the rivals was to firebomb the opposition, leaving gutted shells of buildings considered to be almost worthless. In addition to these places, there were many vacant houses in Woolloomooloo as a result of resumption for the construction of the proposed Eastern Suburbs railway and expressway. Soon there was a large population of derelict men in the area who quickly occupied any vacant house and, sadly, often lit newspaper fires on the floor in winter and incinerated themselves as well as the house.

As I roamed around these streets, looking for properties to buy, Alf was my constant companion and bodyguard. He could sniff out a derelict a block away and we had no problems taking possession of the vacant houses we bought. Unfortunately the family doctor also sometimes smelt of medical alcohol and Alf did not discriminate. Occasionally I was glad of Alf's company too when I ran into my criminal neighbours. Most of them were okay about me, but as I expanded my property portfolio a couple became antagonistic. I remember one confrontation in an East Sydney back lane when the message was delivered: 'I'll get you with a fuckin' iron bar, mate. I've done it before!' Alf growled, but he was as keen as me to get to a major thoroughfare.

Fortunately the heat went out of the gang wars fairly quickly, but at their peak, with car bombings commonplace and threats being made on a daily basis, I took to sprinkling sawdust on Judy's car bonnet each evening, so that in the morning I could see if it had been tampered with.

With the properties bringing in good rents and IBK Constructions roaring along, I turned to my neglected love, sailing. It hadn't been easy, giving up all but the occasional Sunday sail, but I felt the sacrifice had been worth it and now I could afford a boat of my own. I purchased existing plans for a 30-footer from the well-respected naval architect Warwick Hood for $200, which I considered exorbitant but paid anyway. Then I started building the plywood chine sloop in the backyard of our house in Superba Parade, Mosman. It was a mammoth project which filled my nights and weekends for well over a year, giving Judy, who was pregnant again, and our neighbours plenty of grounds for noise complaint, but I delighted in seeing this beautiful thing taking shape before me.

In the winter of 1969 the sloop was nearing completion. I was looking for someone to paint her and when I enquired around the waterfront at Mosman Bay I was pointed in the direction of a rather aloof fellow we knew alternatively as 'The Bilge Rat' or 'Palladin', after the Richard Boone 'Have Gun Will Travel' character he strongly resembled. His real name was Peter Livingstone Curran, a Tasmanian. He was a gypsy character who'd come to Sydney to work as a builder on the Menzies Hotel, then hopped a ferry for Manly one day and wound up in Mosman Bay by mistake. He said, 'This'll do me,' and on and off for the rest of his life it did. I hired the Bilge Rat to paint the hull and in time he became one of my closest friends.

I named the boat *Bimblegumbie*, an Aboriginal word meaning 'the wind that comes from the woomera', and made preparations to launch her. Judy was to do the honours but on the big morning she woke up before dawn and burst into tears. I said: 'What's wrong!'

'I've got to go to hospital. I'm in labour!'

I drove like a maniac to the hospital where Judy was whisked away to have the baby in private, as was generally the custom in those days. While I waited nervously, the sister periodically stuck her head around the door and called cheerfully: 'Won't be long now.'

After she'd done this once too often, I called back: 'For Christ's sake tell her to hurry up. I'm paying waiting time on the crane!'

Our beautiful daughter Pip was born in her own sweet time and I raced back to Superba Parade, where Judy's father, Tom Allen, was running the show, refusing to let them move the boat until my return. The boat was slung onto a low-loader and taken to Rose Bay, where our friend Barbara Mendoza smashed the champagne on Judy's behalf. Later the Bilge Rat and I had her towed back to Mosman Bay to start fitting her out.

This was another job that took all my spare time for months. One day I was working on the bare hull when a big, rough, tough bloke came ambling along the wharf carrying a heavy sugar bag, which he handed to me on the deck of *Bimblegumbie*. 'Here ya go, mate. Might be able to use these.' Inside the bag was a set of cockpit winches. The bloke was Syd Fischer, a great skipper I later sailed with on *Ragamuffin*.

We nearly lost *Bimblegumbie* before we even sailed her. One night I went down to the boat after work and she nearly toppled over in the water as I climbed aboard. The Bilge Rat was nowhere to be seen, so I walked up to our drinking establishment, the Mosman Rowers, and found him at the bar. 'Have you been on the boat today?'

'Yeah.'

'Was she a bit lively?'

'Nah.'

'Well, she is now!'

'Keel's probably fallen off.'

I had a beer in front of me by this, but the Bilge Rat's words, whether meant in jest or not, sent a shiver down my spine. I put my drink down and hurried back to the boat. I stood on deck and ran the boat hook down the bow and along the hull. The keel was there all right, but as I ran the hook back along it, it suddenly shot up. I whipped down below and ripped up some floorboards. All the keel bolts were still there but I picked at one and plucked it like a carrot out of moist ground. Water gushed in. I ran back up to the bar and screamed: 'The bloody boat's sinking!'

The Bilge Rat said: 'Told ya.'

We grabbed as many corks as we could find, plugged the hole and bailed her out. Next day we slipped the boat and bolted the keel back on. It was electrolysis in the stainless steel, like white ants in timber. Could have happened to anyone, but it happened to me with my first boat. I was the butt of a good few jokes around the waterfront, that I didn't fully appreciate!

Bimblegumbie was the best boat I could afford, but not quite the boat I wanted. That much became obvious soon after I started sailing her. We had friends who had a holiday house on Bowen Island at Jervis Bay on the NSW south coast and we had been invited to spend some time with them, so Judy and the girls drove down the Princes Highway while I sailed down with a couple of mates. We roared off down the coast, no life raft, no life-jackets, no radio, no worries. After the holiday I sailed back with another mate and we ran into bad weather. I was relatively inexperienced at this kind of sailing, but even given that, I knew *Bimblegumbie* wasn't handling the going.

When I got back to Mosman Bay, I said to the Bilge Rat: 'I want to be the weakest link in the chain. The boat has to be stronger than me, and this one isn't.'

A short time later I had another close squeak in her. I'd spent the weekend up on the Hawkesbury River with a mate and was coming back on the Sunday night when a gale blew up. The seas were horrendous and we fell off the back of a big wave and split the plywood hull right down the side. We pumped and bailed all night to prevent her from sinking and limped, exhausted, through the Heads next morning.

The Bilge Rat shook his head when he saw the damage.

'I know,' I said. 'She's gotta go.'

CHAPTER FOUR

MEETING MARIS

'Ian Kiernan drives around the streets of Surry Hills, Woolloomooloo and East Sydney like a man obsessed.'

So began an article in *The Australian* newspaper in August 1969, just a month after the birth of Pip and the launch of *Bimblegumbie*. It continued:

> What he sees from the wheel of his battered utility is a 'magical' world of desirable residential property. His stake in this tenement belt — 'brothels, rats, bomb blastings and all' — is large. Already he owns 30 houses, has another eight under renovation and intends to go on until he owns at least 100.
>
> Ian is a new breed tycoon. He shows a certain contempt for the usual accoutrements of success. His office is spartan. He wears blue jeans 'with the bum out of them', a T-shirt and sloppy cardigan. His dog sleeps under the desk. Within a three-quarter mile radius of his office he has 25 men working for him, reshaping and revamping little old houses which are gradually but perceptively changing the face of the area.

The article went on to explain what IBK was doing in the old slum belt and led into it with the first of many press photos of me with

Alf, the wonder dog, plus a couple of before-and-after shots of IBK's current miracle restoration. Suddenly IBK was the saviour of the slums and I was the pin-up boy builder. Well, sort of . . . I tried not to get too carried away by our success.

IBK expanded into a whole fleet of companies as I bought up properties, but inevitably it reached a position where it could go no further without an injection of funds. I had been carrying out renovation contracts for the senior partners in the legal firm Gaden Bowen & Stewart, and Gordon Stewart had become a friend. Gordon had been a speculator on inner-city properties for some time, using me as a renovator. We both knew the value of the inner city.

Finally, at a meeting with Gordon I put my cards on the table: 'Mate, I'm totally geared and I can't raise another buck, but all around me I see opportunities I can't let go.'

Gordon said: 'I can see them too, and I've got access to finance so let's see if we can do some deals.'

> *I knew he was a very efficient operator in the building business. He'd done some renovation work for my properties in Double Bay and he had the sort of taste I approved of. We were really in the business of making desirable residences out of rubbish heaps. These were the days of high inflation and low interest rates. You could buy a terrace house for $15,000 and make a net return of 10 per cent without doing anything. If your interest rate was 8 per cent you owned it for nothing with full gearing. That was the name of the game.*
>
> Gordon Stewart, lawyer and businessman.

We cut a joint venture deal very quickly, with IBK putting in the skills and Gordon's investment company putting in the new capital. Our first joint venture was the purchase of 136 Liverpool Street, Darlinghurst, for $13,000 from the estate of Joe Borg, a gangster who was car-bombed in 1968. As East Sydney's top pimp he was said to have made $10,000 a week, but when he died he left only $65,000, all of it bequeathed to the RSPCA in what may have been his only unselfish act. This property was a corner block which was to become the foundation stone of a development known as Squizzy Taylor Square. But that came later.

In the meantime Gordon and I started to buy wherever we could at the right price, working on the sailing tactic of 'loose cover'. We sought to establish a loose cover over the East Sydney Basin, buying up key parcels of land that we could earn income from and establish a strategic hold for future development. But there was one crucial difference between us and other developers. Our policy was interim development, which basically meant the re-use of existing buildings, pending a city plan that encompassed the protection of heritage buildings. By 1970 the speculators had moved into East Sydney and Woolloomooloo in a big way in anticipation of the release of the State Planning Authority's rezoning scheme for the area. While this was good for the values of our existing properties, it raised the spectre of the area becoming a high-rise concrete jungle. Sir Robert Askin's government had a ludicrous plan for 9:1 floor space ratio development which would make skyscrapers almost mandatory. As Sydney's *Sun* newspaper reported: 'A new city, where property values are sky-rocketing each week, is about to be born within Sydney . . .Wealthy speculators, anticipating that the new (SPA) scheme will permit high-rise development, have already moved into Woolloomooloo, sending property values beyond imagination.'

By now IBK's restoration work was respected enough for me to be invited to join the Historic Building Advisory Committee of the Sydney City Council, and to work with the National Trust and others to save historic buildings earmarked for demolition. I was a developer and I was in East Sydney for profit, make no mistake about that, but I had — and still have — a genuine love of the character and style of the old inner city. However, as opposition to the proposed high-rise development increased, and the Builders Labourers' Federation geared up for a decade-long battle, I found myself tarred with the same brush as the multi-millionaire profiteers who wanted to blot out the sun from Woolloomooloo's streets.

Meanwhile, Gordon and I continued our buying spree. We bought three properties in Crown Street and another three with commercial zoning in Cathedral Street. Each block cost $20,000 and we invested another $20,000 in them. IBK now stopped outside contracting and worked full time restoring our own properties to prepare them for renting. I had become a regular fixture at Sydney Town Hall where every other day I would appear with plans for approval. I had a great rapport with a bloke named John McClements who became known as 'the IBK clerk'. We were taking out first mortgages at 7.5 per cent and second mortgages from about 11 per cent while our rental returns were providing 18 to 20 per cent on investment. After the places had been let and the rents started to creep up, we would revalue and refinance them. In fairly quick time I found myself a director of a dozen or more companies and a millionaire, on paper at least. I bought myself a Porsche and Judy a red Mercedes and set up a city office, first in Palmer Street and later at 159 Cathedral Street.

Around the time I started doing joint ventures with Gordon Stewart, Judy and I were invited to a cocktail party at the Mosman home of the managing director of Reckitt & Coleman, who was a friend of Judy's. There we met a tall, rangy Englishman named Peter Mendoza and his wife Barbara, a stunning woman who had been a fashion model in Europe. The four of us got on well from the start, and later, over dinner, Peter explained that they had sold their property in England and were looking to invest in Sydney real estate. The couple asked me to help them find a property suitable for renovation. I found a waterfront place in Bay Street, Greenwich, purchased it for them at a good price and put a second storey on it. They were impressed and asked me to enter into a joint-venture program, along the lines of my arrangements with Gordon. While I set the new business in motion, the Kiernans and the Mendozas became firm friends, so much so that I asked Barbara to deputise for Judy at the launching of *Bimblegumbie.*

For our first venture we bought a couple of houses in Crown Street, Surry Hills, which I sold on almost immediately for a good profit. Then we bought terraces in Reservoir Street, Goodchap Street, and a row of houses in Corben Street, behind the terrace house that was to become Sydney's plushest brothel, the Touch of Class. By 1970, with both joint ventures flourishing, there was barely a street in East Sydney without at least one sign which read: 'This building is being renovated and faithfully restored by IBK Constructions.'

In late 1970 I had a marvellous afternoon out on the harbour on *Bimblegumbie*, with a stiff nor'east sea breeze filling her sails. I felt totally invigorated when I brought her back into Mosman Bay, but the sight of a yawl newly anchored in the bay brought me unstuck. I gaped at this beautiful craft for about twenty minutes before I made my way up to the Mosman Rowers to have a beer with the Bilge Rat.

'Know who owns that yawl?' I asked him.

'That's *Maris*, Jack Earl's boat. Been cruising in the Pacific for a few years.'

I knew who Jack Earl was. Everyone on the waterfront did. He was a legend. Born in Durban, South Africa, in 1908, Jack had arrived in Australia in 1911 and spent much of his childhood on remote Darnley Island in Torres Strait, where his father was the schoolteacher. Here he learnt to sail with the Islander children in their dugout canoes with sapling masts and bed-sheet sails. When he finished school in Sydney he became a newspaper artist by profession and a sailor by inclination, saving every penny he could to buy a succession of boats, and eventually commission a 44-foot gaff cutter he named after his wife, Kathleen Gillett as she had been born. The boat was launched in 1939 and the Earls planned to sail her around the world, but war intervened and Jack joined the Naval Auxiliary Patrol.

By the end of the war the Earls and their two children were living on board the *Kathleen Gillett* at an anchorage just inside North Head. To celebrate the peace, Jack had decided to spend

Christmas sailing down to Tasmania, so one afternoon he rowed his dinghy over to *Saltair*, a boat owned by Bert Walker, a Tasmanian who had extensive charts of Bass Strait and the Tasmanian coast. As the two men pored over the charts and consumed a bottle of rum, they decided to sail south together. Soon there was a small fleet preparing and Captain John Illingworth, an Englishman with the Royal Navy stationed in Sydney and a noted ocean racer, suggested they make a race of it. This was the birth of one of the world's great ocean yacht races, the Sydney to Hobart.

In 1947 and '48, Jack finally sailed the *Kathleen Gillett* around the world and she became the first Australian yacht to complete the circumnavigation. In the 1950s he became a full-time marine artist and commissioned another boat from *Gretel II* designer Alan Payne, a 36-foot Huon pine yawl to be built in Hobart by Jock Muir. He named her *Maris*, after his daughter. This was the boat I fell in love with at first sight twelve years later and still love today.

Jack Earl had slipped the boat on his return from the Pacific and the Bilge Rat had done some work on her. He said: 'Jack's gettin' on a bit, mate. If you're interested I reckon he'd listen to an offer.'

In fact, it transpired that Jack had already listed *Maris* with a broker, but a few days later he had taken her off again. He was prepared to sell her, but he regarded a relationship with a boat as a very personal thing and couldn't bear to see it reduced to mere commerce. The Bilge Rat conveyed my interest and Jack suggested a meeting. We met on board and I was again impressed by the fine lines and robust construction of this little boat. I was equally impressed with her skipper. Jack Earl was then a little past sixty, a bloke of average height, balding and sporting a beautifully waxed moustache. He had a style about him which combined elements of the bohemian, the Pacific Islands cruising skipper and the waterfront knockabout. He loved a drink and a laugh, and we got on like a house on fire.

That afternoon we sat in the cockpit of *Maris* on Mosman Bay and thrashed out a deal over a few rums. In the middle of it a southerly buster came roaring up the bay as a Bluebird sailed towards us with a fat man standing on the deck. Fifty knots of breeze picked her up, the Bluebird did a massive Chinese jibe and the boom knocked the fat man into the water. He started screaming that he couldn't swim so I dived, fully clothed, into the Harbour and dragged him out. Together we hauled the fellow up on deck and he sat there like a huge wet seal. Jack thought the episode was hilarious.

The agreement to buy *Maris* was no ordinary deal. It involved much more than money. In fact, although Jack was by no means rich, money was the least important part of it. He had to be convinced that I was the right sort of fellow to take on the responsibility of caring for his boat. It was like parting with a good dog. It doesn't hurt so much if you know the future will be okay. Anyway, Jack ended up with some money, a new shed which I built at the Earl 'compound' overlooking Mosman Bay, and a new mate. I ended up with *Maris*, a beautiful painting of her, lessons in sailing and celestial navigation, and the finest mate a man could wish to have.

I started spending time at the Earl compound and got to know Kath Earl, a beautiful and wise woman, their daughter Maris, and their extended family of grandchildren. Jack's grandson Ben Hawke joined me on the boat for a Christmas cruise to Port Stephens.

During the negotiations to buy *Maris* I'd met Don Mickleborough, a veteran sailor and incredible character who owned the famous 38-footer *Southerly*. The exploits of Don and his crew were legend around the waterfront, so when he invited me to sail to Port Stephens to join him at Christmas, I was delighted.

The following Easter both Jack and Ben were with me when I made my first ocean crossing, the short run to Lord Howe Island. By then I'd learnt the basics of celestial navigation and it was wonderful to be at sea with a real master of the science. They were great shipmates, Jack and Ben; one with the knowledge of the ages, the other with the exuberance of youth, both with an affinity with the sea that only a few sailors attain.

That trip — we called it the 'delivery' trip, despite the fact that I'd owned *Maris* for four months — also introduced me to a new love. We sailed into the shallow, horseshoe lagoon at Lord Howe Island on a beautiful clear morning with a breeze pushing tufts of white cloud around the twin mountains, Gower and Lidgbird. The water was crystal-clear and you could see the schools of fish scuttling away as our bow sliced through the water. Along the shore there were a few houses visible amongst the trees, but overall the impression was of an untouched paradise. A quarter of a century on, I have sailed into a thousand sun-sparkling lagoons, and not one of them has even come close to matching the beauty of Lord Howe, a place where I've made innumerable friends.

We stayed on the island for a couple of weeks, then Jack flew home and the Bilge Rat and Dick Logan from the *Southerly* crew flew out to help me sail *Maris* back. I was thrilled to be making the return journey as both master and navigator of my own vessel, but I didn't yet know enough about sailing by the stars to do it without Jack. We ran back to Sydney on dead reckoning, a form of navigation in which you use the boat's speed, compass heading and estimates of leeway and drift to calculate a reckoned position. It got us there and in the last hours a sixty-knot southerly buster welcomed us home.

With IBK and the joint ventures rolling along nicely, I made time that winter to get to know my boat. By July, Jack had taught me to navigate by the sun, the moon and the stars, I felt I was extracting maximum performance from the little yawl and I was keen to make another voyage. I planned to sail to Queensland, but business got in the way and I had to cut it back to Coffs Harbour. I put a crew together consisting of Bob Lane, who was a family friend, Soren Lueno, who had Soren's Bistro in Woolloomooloo, and a Danish mate of Soren's. We had gusty westerly winds right up the coast and made good time. The trip was a beauty, except for Soren's mate who was hit by *mal de mer* and spent most of his time lying on a bunk and throwing up all over the cabin floor. A couple of Tasmanian sailing mates had told me that their Bass Strait solution to this problem was to put oilskins on the offender, attach him to a couple of lifelines and tie him to the afterdeck. That was what we did with this bloke. I don't know that it made him feel any better, but it was a lot more pleasant below.

At that time there was only an outer breakwater at Coffs Harbour, so we anchored inside it with a couple of other boats. We partied pretty hard that night, 25 July, and didn't realise that the low pressure cell which had been providing us with the westerlies had now intensified. If I had realised the wind was gusting to sixty-five knots I probably would not have been worried too much anyway. I had good ground tackle set with a twenty-two-kilo 'fisherman's style' anchor, six fathoms of forty-millimetre chain and heavy rope. With my limited experience that seemed like more than enough to secure the boat.

In the early hours of the morning I was sleeping off the party below deck on a yacht called *Tara* when I heard my name being called. I drifted in and out of sleep, then suddenly realised the call had an urgency about it. I raced on deck and saw Bob Lane standing in the stern of a fisherman's dinghy in the driving rain. 'Come on!' he yelled. 'The boat's on the beach!'

I jumped on board, my heart racing. How could this have happened? Bob started to fill me in. They had been sound asleep when *Maris* slipped her anchor and almost sailed out of the harbour, pushed seaward by the howling wind. But she had clipped the south side of the breakwater and the two-metre concrete cubes had ripped a huge hole in her port side. Awoken by the tremendous crunch, Bob and Soren saw that she was taking water and decided to get the boat to the beach. They'd had difficulties starting the motor and the damage had increased. Now my beautiful *Maris* was awash in two metres of water, wrecked through my own stupid neglect. I retrieved the brass compass binnacle as it rolled out of the cockpit and let out a roar of frustration and anger.

As first light came into the winter sky, I saw a man walking his two greyhounds along the beach. Seeing our predicament he hurried off and returned with a Land Rover. We got a line on the bow and managed to haul *Maris* into shallower water. The work was arduous and we were wet and cold. I remember taking a break and lying on the sand, shivering in my wetsuit and feeling like a whipped dog. Then a couple of fishermen arrived with more bad news.

'Lock her up, son,' one of them said. 'The roll back is coming and she'll fill with sand.' I'd never heard of roll back. They explained that in a fierce westerly storm the sea humped up on an offshore plateau, then came rolling back as breaking surf. As I battened down *Maris* as best I could, I kept repeating a mantra to myself, 'I will not be beaten, I will not be beaten . . .'

The local newsagent took pity on us and let me use his phone while his wife made us a hot meal. My first call was to Gordon Marr, my old workmate at G.J. Coles, whose family had a haulage and heavy crane business in Sydney. I said: 'Gordon, send me your best bloke. There'll be a ticket waiting for him at the airport.' Then I phoned Cec Quilkey, a talented wooden boat builder from the south side of Sydney who had built many boats, including Syd Fischer's famous Admiral's Cup yacht *Ragamuffin.* Cec was on his way the moment he put the phone down.

The salvage mission began. Ralph Featherstone, a school mate from TAS arrived to help, and a local crayfish diver offered a hand. I asked his name. 'Call me Shark Shit,' he announced cheerfully. 'That's what they reckon I'll end up as.' We borrowed a couple of cranes and a bulldozer from a timber camp. The harbourmaster loaned us a massive mooring line from a Japanese fishing boat. Unfortunately the cranes couldn't get close enough as the roll back had started in earnest and was dumping three-metre waves onto *Maris,* blowing the forward hatch and shooting sailbags high into the air as the surf roared through the boat. The bulldozer strained as its tracks turned in the soft sand.

We worked all day. The rope broke three or four times and we would have to brace ourselves against the bulkhead and wait for a break in the waves before we could retie it. As dusk approached

Maris was three-quarters filled with wet sand. The bulldozer driver said: 'Matey, unless you get her out with this wire rope, she won't last another high tide.'

We passed the wire rope around her and through the propeller aperture. The bulldozer winched from the other side of the sandhill to get purchase. *Maris* finally started to move slowly up the beach. I was glad to be in the breaking spray so that no one could see my eyes fill with tears of relief as she inched up to the high tide line.

I went out to the airport to meet Marr's rigger. I got the shock of my life when Gordon Marr himself stepped off the plane. 'You said you wanted my best man,' he grinned. I knew with that sort of spirit we would save *Maris*, and that night, under the lights of four-wheel drive vehicles, Cec Quilkey pronounced her repairable.

At first light next day we were on the beach shovelling sand out of the hull. There must have been seven or eight tonnes of it and our fingernails were soon stripped to the quick from scraping it out of the tight corners. Gordon scrounged four back-end cranes from local railway sidings and we bought enough hardwood to pave the beach so that the rubber tyre vehicles could get alongside *Maris*. We lifted her as the low-loader backed slowly underneath her damaged hull. Then we took her to the jetty and washed the rest of the sand out with fire hoses. That night she was snug down on the low-loader and bound for Cec Quilkey's boatyard in Taren Point.

We lit a huge bonfire on the beach to get rid of all the hardwood while we toasted the success of the rescue operation with rum. The next day, when they noticed that *Maris* was gone and there'd been a fire on the beach, the locals gave birth to a rumour that persisted for years — that we'd burnt the boat. A small article appeared in the Sydney *Daily Telegraph*:

> HOLED YACHT SALVAGED
> The $24,000 Sydney-owned 36-foot yawl *Maris*, wrecked in a storm on Sunday will be taken to Sydney for repairs. Owner Mr I. Kiernan today stripped the yacht of its masts, engine and other movable parts, then loaded it on road

transport for Sydney. The yacht's bow was torn open and the keel almost ripped from the hull during salvage operations.

When I got home to Mosman I was still shaking from the episode. The immense relief I felt at having been able to salvage *Maris* was tempered by the huge embarrassment I also felt. I had to tell someone who'd understand, so I found the Bilge Rat and invited him home for a meal. We went to the fish markets and bought a live mud crab for the occasion. At home I boiled up a pot of water and cut the strings to drop him in. The crab reared up. 'Look at him,' I said. 'He knows we've got to kill him but he won't accept it. This is one courageous crab!' I suppose we'd had a few drinks by this, but it made perfect sense to us both. We got in the car and drove down to Mosman Bay where we slipped that brave crab back into Sydney Harbour, then went home to eat rice.

Jack and Kath Earl had gone to Canada and California for the winter but I had to let them know about *Maris*. I wrote to him at an address in Sausalito and soon received a reply. 'Your troubled face has been constantly before me and I hope by this everything has settled . . . Congratulations, Ian, on your tenacity and the beauty of your thought through this tremendous effort.'

I didn't think I deserved to be congratulated, but Jack always knew how to lift my spirits, and this was the first of many letters I was to receive in a hundred ports over almost twenty-five years. Not once did I open a Jack Earl letter and not find it filled with love, laughter and hope.

CHAPTER FIVE

THE DUKE OF WESTMINSTER

In 1969 I had bought the old Federal Broom Company building at 200 Crown Street, Darlinghurst, not really knowing what I'd do with the fire-gutted wreck. But, as was my custom, I put up a big sign announcing: 'This building is being renovated and faithfully restored by IBK Constructions as two levels of prestige office space and a restaurant.' I was working on the site a couple of days later when a woman walked in and started gushing in a thick French accent: 'Zis is fantastic, I have dreamed zis only last night! I want to make here a fabulous restaurant, a special restaurant. I will call it *La Rive Gauche.*'

Her name was Monique George and she signed a lease that day. I fitted out the restaurant for her and it became one of Sydney's finest dining establishments. It also gave me a taste for fitting out restaurants.

At that time I was running and swimming at Balmoral every morning and I would often see a big, grey-bearded fellow sitting outside the bathers' pavilion, sipping coffee and staring out to sea. I pointed him out to Judy one morning and she said, 'Oh, that's Mischa Kanaef. Used to be in the film game. I know him a bit.

Nice chap.' Some weeks later Mischa's name came up again. Judy said, 'You've got to talk to him. He's in trouble and you might be able to help.'

Mischa came to see me and told me his story. He too had a dream to build a beautiful restaurant on Balmoral Beach. He'd secured a lease on the bathers' pavilion from the council for $65 a week, engaged architects and engineers to plan the renovation and then speculated on the stock market to raise the money. He'd failed miserably as a share punter and now he was broke, living in the pavilion, and selling chips and cold drinks at the little kiosk to keep the wolf from the door.

I told Mischa that as much as I sympathised with his predicament, my game was buying property, not leaseholds. The meeting ended, but next day he phoned me again. 'I know what you said yesterday but I've written down five very good reasons why you should support me,' he said. 'May I show them to you?' We had another meeting and I read what he'd written, then I looked at him. He reminded me of the mud crab the Bilge Rat and I had freed, only bigger. He was beaten but he just wouldn't accept it. If I have a weakness in my judgment of others (and I'm sure I have many), it is that I admire that dogged determination to win to the point where it sometimes blots out other considerations. Perhaps this was why I told Mischa I'd think about his deal.

I went to Gordon Stewart with Mischa's scrap of paper and we crunched the numbers. Finally, Gordon said: 'Why don't we do it?' The first step was to bail Mischa out, to get the creditors off his back. Once we'd done that I got the approvals and built his restaurant, which was a huge and immediate success, of which Gordon and I owned fifty per cent.

Johnnie Walker, who had the famous steak house in the city, wrote the first review:

> This is the story of a dream come true, and it's of more than just human interest because it seems at last Sydney has the sort of seafood restaurant it deserves . . . Mischa's dream

> came true just before Christmas when he opened Mischa's in the old bathing sheds at Balmoral Beach . . . For a start, the place looks superb. It is built looking out over the water and three huge arches afford generous views of the Harbour outside; it is like something you would find in southern Spain . . . I would say he has got just about the best designed restaurant kitchen in Sydney . . .

And he said the food wasn't bad either!

Unfortunately, the dream soured for me. Mischa and I fell out and Gordon and I eventually sold our shares, but the venture had shown me that there was good money to be made in the restaurant trade, and I started several others over the next few years.

Since our initial purchase of 136 Crown Street, Gordon and I had bought up much of the adjoining property. We now had virtually all of what amounted to three city blocks, bounded by Liverpool, Crown, Burton and Palmer Streets and intersected by two lanes. Since most of the properties had commercial zoning, I had the idea of removing all the back fences and putting together a kind of 'business village', a $2-million development which would be the biggest gamble in our joint-venture history, but which I believed had the potential to attract the 'trendy' businesses — publishing, advertising, public relations, architects, designers — who would be prepared to pay a premium for access to a pleasant and different environment. We drew up plans showing terraced offices with common courtyards shaded by trees.

One of the first businesses to sign a lease was Singleton Palmer Strauss and McAllan (SPASM), an advertising agency headed by a young man named John Singleton. Singo and I became mates and he quickly pointed out that this was such a big project that I needed to communicate what we were doing. He put me in touch with one of his associates, a public relations consultant named Pauline McGrath, who immediately set about raising our profile in the media. But Singo's greatest contribution was to come up with a name for the development. We'd been calling it the Crown

Street Commercial Village, which was about as exciting as pie and peas. For contrary reasons Singo suggested we name it Squizzy Taylor Square, after a Melbourne gangster who'd never even been to Sydney. It wasn't as tasteless as calling it 'Joe Borg Square', but it had the right crime connotations. There was a public outcry, of course, and soon everyone in Sydney knew about Squizzy Taylor Square. Then Singo delivered the *coup de grâce.* We took out full-page ads in the newspapers with a picture of gangsters wielding machine guns. The bold type read: 'Rent some space or we'll kill you!' It was vintage Singo.

The rest of the message was slightly more subtle: 'Let's say you're famous. An architect or something. And you're a little bit tired of the inner-city ploy. And you want to live your business life much the same as you spend your leisure time. OK. The following may be of interest. You see, we've set out to create a working environment which is utterly conducive to the new, relaxed style of business in the seventies . . .' It worked brilliantly. Soon Squizzy Taylor Square was full of Sydney's 'beautiful people', happily paying $4.75 a square foot for space within our 40,000 square feet. Of course we also had some of Sydney's least beautiful people within the precinct. The criminal element was still very much a part of East Sydney life, and one of their favoured meeting places was the Maltese Club, a building right in the middle of Squizzy Taylor Square that we had been unable to buy. After I moved into the Square the fellow who ran the club and I developed a nodding and watching relationship. He was no real trouble but I would have liked to see him go. For his part, he seemed to be waiting for me to offer him a large bag of money to go away.

Life was starting to be very kind to Ian Kiernan. I was just thirty-one and in control of a property empire worth close to $15 million. I was sensibly geared and the variety of investments in my property portfolio meant that I had a degree of protection from economic downturn. Looking back now with the dubious benefit of hindsight, I can see that this was perhaps when I should have got up from the card table. But in the early '70s I was a young man who romped around town in a leather jacket — often astride a new Honda 4 motorcycle — and could do no wrong, at least not in business. I played on and brought my lifestyle more into line with my new turnover. Judy and I bought a three-storey house in Moruben Road around the corner from our bungalow at Balmoral and we began to restore it as the new family home.

Around this time the Bilge Rat and some others around the waterfront started referring to our new home as 'the castle'. It wasn't long before I was 'king of the castle'. In the curious way that nicknames evolve in Australia, I went from 'king' to 'King Biscuit' (after a rock band) and finally to just 'Bicky'. The last name stuck and I have been Bicky ever since.

Despite all our purchases, I wasn't into conspicuous consumption for its own sake. I'd just made a conscious decision that if we could afford it, we would make no compromises in quality. That policy applied to the rebuilding of *Maris* too. I supplemented the insurance money to ensure that she was fitted with the best of everything. No expense was spared in locating the very best Huon pine, the best boat-building timber in the world and increasingly hard to find. Hood Sailmakers were commissioned to supply a new wardrobe of sails. Little *Maris* would once again be the most beautiful girl in the bay. I would make sure of that.

While the rebuilding was taking place at Cec Quilkey's yard, I would make the time to visit *Maris* once or twice a week. It was wonderful to see her coming back to life. When Jack and Kath came back from California I sometimes took them with me, making an outing of it. We'd stop at Botany Bay and eat our lunch at the picnic tables, then spend the afternoon examining the repair work.

With *Maris* out of action for the summer, I bought myself a second-hand Hobie Cat so that I'd have something to sail, but then a much greater opportunity presented itself. The maxi yachts were arriving from America to compete in the Southern Cross Cup, one of Australia's greatest sailing regattas. Jim Kilroy's *Kialoa II*, Ted Turner's *American Eagle* and Hughie Long's *Ondine III* were some of the biggest name boats to arrive. Don Mickleborough knew all these guys, and through his introduction and influence I was invited to crew on *Ondine.*

I was absolutely thrilled. I had never raced a maxi and here was a chance to broaden my ocean racing experience and to learn how to handle big gear on a big boat. The crew was big too, as befits a 73-footer. We were twenty-two in all, and my shipmates were a great bunch. Nick Hilton was the sailing master, a big blond bear of a man. Welshman Peter Morgan, New Zealander Sid Brown and Californians Ken Gardener, John Piggott and Terry LeLong were just some of the others I got close to, and whose dry sense of humour I came to appreciate.

The final race of the Southern Cross Cup is the Sydney to Hobart, in which we finished second to *Kialoa.* Then came the festive season in Hobart, which featured some extraordinary scenes, not the least of them being the nude wedding of one of the competitors on board an old ferry on the Derwent.

After the regatta *Ondine* was heading home by way of New Zealand, but many of the crew had already flown out, so I decided to crew the first leg to Wellington. Nick Hilton was the skipper and Peter Morgan and myself watch captains. Also on board was an attractive girl known as 'the tree lady'. A couple of months earlier at a barbeque on Store Beach, Don Mickleborough had pointed out an attractive, scantily-clad girl perched in a tree. Nick had coaxed her down and later signed her on as cook. They eventually married and settled in New Zealand, and she is still his 'tree lady'.

The voyage to New Zealand was a dream. We made good time crossing the Tasman and in one twenty-four hour period ran 273 miles (437 kilometres), averaging 11.5 knots. That year the Tasman seemed to be full of whales and we sailed through pod after pod. Some came so close you could almost reach out and touch them. The species we saw varied from the giant sperm whales to the little black whales and medium-sized humpbacks. Many had calves swimming along beside them. These and the

Southern Ocean seabirds, the albatross, petrel and fulmar, had me fascinated. It was as well I enjoyed the sailing because there was plenty of it. Nick spent a lot of the voyage down below in the bunk with his tree lady.

When I got back to Sydney from Wellington there was a message on my desk to call Gary Bogard. Gary was a bright young lawyer at Freehill Hollingdale & Page who'd gone into partnership with a man called Sid King in a development company they called Home Units Australia. I'd been put in touch with him the previous year by real estate agent Spencer Simmons. Simmons had phoned one day and said: 'I think I might have a buyer for your Cathedral Street properties.' This was a bit surprising since they weren't for sale, but you learn quite quickly in property that everything is for sale. Gordon Stewart and I sold Cathedral Street and walked away with $80,000 profit, which we put straight back into other purchases in Woolloomooloo, one of which was a rundown old pub that was let out as a boarding house for derelicts but was still known as the Globe Hotel.

I phoned Gary Bogard. He got straight to the point. 'We want to buy the Globe Hotel.' I told him it wasn't for sale, but he arranged for me to have lunch with him and Sid King anyway. 'Just a social lunch, a friendly chat.' We were to meet at Fred's, a restaurant in Woolloomooloo, but for some reason I was seconded to pick up Sid King on the way. We'd never met but he was as I expected, a flashily dressed little chap. Alf took an instant dislike to him and an insatiable lust for the cuff of his suit. I had to disentangle them and throw Alf in the back of the ute. He continued to growl at King from under the table throughout our lunch.

During the lunch I made the point several times that I wasn't a seller. Finally I said: 'Look, I don't want to sell the Globe Hotel, but I might be interested if I could retain some equity in the venture.'

Sid King chimed in: 'How much do you want? Fifty per cent?' I tried not to gulp. It looked like we were venture partners with Home Units Australia.

Over the next few months I proceeded to negotiate the purchase of the various remaining properties in the block adjacent to the Globe. When these were all secure I asked Spencer Simmons if he had a buyer. Yes, he thought he did. I rang all the directors of our joint venture and explained that Londish Holdings was prepared to buy the proposed development at a price that would realise us a profit of $1.13 million. Should we sell? 'How quickly?' was the rhetorical question with which they replied.

If the joint venture with Home Units Australia was what propelled me into the heavily-geared financial big league, it was Pauline McGrath and John Singleton who turned me into the best-known builder in town. Being something of an extrovert, I didn't mind the attention at all, but it was to have its downside. In time I came to loathe the description 'trendy developer'.

Tony Reeves, a journalist who was very much involved in the battle to preserve the inner city and who later became an alderman on Sydney City Council, wrote in one of the early profile articles on me:

> Sitting behind the wheel of his plush automatic Mercedes [it was Judy's], Ian Kiernan reflected this week on his rapid rise to fortune. Seven years ago he was a $60-a-week builder's labourer. Today he is the co-landlord of about $15 million worth of real estate . . . As the biggest private landowner in Sydney who is now spreading into Melbourne, Mr Kiernan has been called 'Australia's Duke of Westminster' . . .

The purpose of the public relations campaign was to differentiate between what we were doing in inner Sydney and what was being done by other developers, like Stocks and Holdings and Sid Londish. Almost from the beginning, I'd taken to painting the exterior of my terraces black as a kind of IBK signature, and as our 'loose cover' increased, we would fill in the black squares on a wall map of East Sydney. We liked to think that every black square on the map represented another small victory for integrity and quality of life in the city. The Builders Labourers' Federation didn't quite see it like that.

The long-awaited Sydney Strategic Plan had been released and Askin's 9:1 ratio affirmed. This meant that for every square foot of land, developers could build nine square feet of commercial space. It was the green light for massive demolition and a new city filled with skyscrapers. The BLF, a strong and vocal trade union headed by a charismatic Communist named Jack Mundey, pledged themselves to stop the destruction. Mundey had involved his union in a bitter campaign to save Kelly's Bush at Hunter's Hill, and no one was under any illusions about the seriousness of the BLF's intent, now that they looked eastward and began the Save the Loo and Save Victoria Street campaigns.

Mundey and his lieutenants, Joe Owens and Bob Pringle, instituted union 'black bans' on development sites which involved the demolition of low-cost housing. Because the idea was to preserve the environment as well as the social status quo, these soon became known as 'green bans'. The battle to save beautiful Victoria Street, Potts Point, from the clutches of the avaricious developer Frank Theeman became extremely acrimonious and focused public attention on the BLF. Australia was still locked in a Cold War mentality, and the fact that the BLF was run by Communists made them the natural enemy. But as the Victoria Street siege wore on, a good many people began to believe that this battle wasn't about politics, it was about hanging on to a cityscape we loved. Theeman's greed and absolute determination to get his own way eventually worked against him, and people like Mick Fowler, Juanita Neilsen, Wendy Bacon and Mundey became household names as they put their reputations (and their lives) on the line to stop him.

Unfortunately the success of the Victoria Street campaign gave the BLF more political clout than was healthy, and I believe they lost the plot. I don't doubt that the intentions of the leaders — particularly Jack Mundey — were honourable, and I honestly think that we have to thank those men and women for the part they played in creating Sydney as we know it today. But after Victoria Street they'd tasted blood and liked it. They went a little power-crazy and decided to stop all development.

By this time IBK had 120 employees, including apprentices in most of the building trades. I had learnt the importance of planning for tomorrow at Monier, and I had my best tradesmen impart to their apprentices the IBK development philosophy, which was mainly about appreciating the artisans of old and taking pride in working with what they had created, rather than tearing it down and starting anew. All this was having some effect, because IBK was now recognised as the restoration specialist and I was frequently invited to speak at development seminars on alternatives to high-rise development. Sometimes at these meetings I would see the leather-jacketed figure of Joe Owens, sitting in one of the front rows, taking notes and scowling. I wasn't with him, so I was agin him, no matter what I said or did.

My personal battle with the BLF started out relatively low-key. I'd arrive at Squizzy Taylor Square to find 'trendy developer' graffiti spray-painted on the walls. This was annoying enough, but we had plenty of black paint and we could deal with that. Eventually a strike stopped work at Squizzy's, but the biggest headaches they gave me were at another development in Woolloomooloo.

With the profit we had taken in the Globe Hotel deal, Home Units Australia and Gordon and I decided to reinvest in the same vicinity and I bought a large block in Brougham and McElhone Streets. This was just below Kings Cross, with sweeping views of the city skyline to the west. We called in the respected architect Ian McKay and began to plan a major restoration of some significant Georgian and Victorian buildings, as well as the construction of some new brick, timber and glass infill buildings. Since this multi-use commercial space was similar in some respects to Squizzy Taylor Square, we decided to call it Ida Pender Place, after Squizzy's girlfriend, a terrible moll by all accounts.

Ida was a $4 million construction project and, despite our

successes elsewhere, cash flow demanded that we get her up and running without delay. Unfortunately this necessity coincided with the need of the BLF to flex its muscles. IBK became the scapegoat. As they applied the pressure, tempers flared. I grabbed one of their men, Ian Makin, in a pub one night and offered to rearrange his face. There was hostility at the site almost every day. Eventually, largely through the work of Bob Pringle, the BLF was able to infiltrate my ranks and get the support of enough IBK men to give me a strike. I was devastated when my men walked off. These were people I'd worked with for years and there was a great deal of mutual respect, but in the early '70s this was the power of the unions. If I was devastated by their departure, I was flabbergasted when I discovered most of them had walked into new jobs. This wasn't in the rules, but evidently the BLF had thrown away the rule book.

With Ida Pender green-banned and no solution in sight, I decided to walk away from it and clear my head. Some Austrian friends were skiing in the Australian Alps, so I jumped in the Porsche and joined them. After some great skiing we drove to Lake Jindabyne to fish for trout. At this we were less successful, but we did catch one large fish, at a place known locally as Shithouse Bay because of a nearby sewer outlet. My friends insisted that the fish was edible and that I should take it home. I drove back to Sydney and took the fish into the office, where I instructed one of the secretaries to take it to Trades Hall and present it to Bob Pringle.

She found Pringle in his office and delivered my message. 'Mr Pringle, I'm from Ian Kiernan's office and he asked me to give you this trout, fresh from the Snowy.' Pringle apparently muttered something about payola and there was a good deal of laughter from the people in his office, but he took the fish anyway.

A couple of evenings later I ran into Ian Makin at the pub. He looked a little worried as I approached and was clearly relieved that I only wanted to give him a message. I said, 'When you see Bob Pringle ask him if he enjoyed his trout. Tell him it came from the main sewer outlet at Lake Jindabyne.'

Eventually the BLF started to see reason and the Green Bans were relaxed. Ida Pender Place went ahead, although the financial effects of the strike remained serious. I found myself talking in a reasonably civil manner with the head honchos, although they still regarded me as the enemy. At one point when I had bought some properties in Louis and Caroline Streets, Redfern, a BLF delegation came to see me and offered some co-operation in return for my giving them the Redfern places for an Aboriginal housing co-operative. I listened to their problems and we talked through a deal which resulted in IBK employing Aboriginal workers to do up the houses that their own people would live in. I flew to Canberra and got the plan approved by the then Minister for Aboriginal Affairs, Gordon Bryant.

The Redfern Aboriginal housing co-operative was a great idea, but it opened my eyes to how little I understood — how little any of us understood — the social problems faced by urban Aborigines. They felt they had no power to influence their situation, they became increasingly frustrated and the cycle of booze, poverty and domestic violence continued. The same guys who were planting IBK trees in Redfern by day would be ripping them out in a drunken frenzy at night.

After almost a year in Cec Quilkey's yard, *Maris* was finally ready to be relaunched in July 1972. Judy and the kids, Jack and Kath Earl, the Bilge Rat, Cec and a few others were on hand when we slid her into a creek off the Georges River and draped a huge garland of flowers over her bow. She looked wonderful — hull and varnish sparkling in the winter sun, new sails, new teak deck, new hatches, radio and bunks. She was better than ever. As we celebrated, I said to Cec: 'I want you to jump with the parachute you packed. What about a shakedown cruise?' He didn't need to be asked twice, and a week later Cec, Jack, young Ben Hawk and I were on our way to Lord Howe Island.

The winter winds blew hard and *Maris* took a lot of water. Her keelson, her backbone, is a log of Tasmanian oak, twenty-three by seventeen centimetres, and after a hot summer in a tin-roofed shed this had shrunk. We took so much water when one big wave hit us that Jack momentarily feared we had sprung a plank, but we made the island with no real problems. On the way back the seas were even more of a test for *Maris.* The wind was howling out of the west at 60 knots or more and at midnight Jack took his watch with two lifelines securing him in the cockpit. He looked up and thought he saw moonlight behind the cloud, a sign that the weather was clearing. The moonlight turned out to be the feathering crest of a massive wave that rolled *Maris* down with her mast and sails in the water. Below deck, provisions and people were thrown around like corks, but the yacht quickly sprang back up and we were able to laugh at the danger now passed. Nevertheless, we hove to until the dawn and waited for conditions to improve.

Jack, seemingly unshaken by his swamping in the cockpit, sat down below and shared a few rums with Cec Quilkey. 'Matey,' he said after the warming glow of a couple of nips, 'I haven't got much sugar but I want a little boat, a 28-footer . . .' The two men sat at the chart table, tossed about by the waves, and drew up the plans for Jack's last boat, *Smoky Cape.*

CHAPTER SIX

FIVE HUNDRED HOUSES

Although I was up to my ears in the property game, sailing was starting to rule my life. I tried to visit *Maris* every day down at Mosman Bay.

I would take care of the little chores that had to be done, then sit at the chart table down below and record my thoughts in a big blue diary. This became my 'Blue Book' and in it, and its successors, I have recorded my most personal thoughts, my goals and ambitions, my disappointments and my regrets. Many of those thoughts have found their way into this book, helping me to remember exactly how I felt at those crucial moments in my life. But some of them will remain between me and the Blue Book, at least until I fall off the twig.

As well as sailing *Maris* for pleasure, I was invited to join Syd Fischer's crew on *Ragamuffin*, the gun boat of the time. This was a tremendous thrill because the *Ragamuffin* crew were recognised on the waterfront as the best in the country, talented and gutsy sailors and real team players who put their all into every race. Those guys became firm friends, as did rough, tough Syd Fischer who, unlike some of the rich men who owned racing yachts, was as good a sailor as the next man on his boat.

In 1972 we did the whole calendar on *Ragamuffin* — the Montague Island race, the Cabbage Tree Island, the fifty-kilometre overnight races . . . every week a new challenge. I probably would have stayed on for the summer and done the Sydney to Hobart on her too, except that I had concocted another plan for adventure. I had decided to sail *Maris* to New Zealand during the summer holidays, and that meant putting my head down and bum up for a couple of months at IBK.

This was not at all easy, because I had further complicated my life by becoming obsessive about motorcycles. My Honda 4 was my favourite land toy and I fell in with a group of guys who used to meet at former dirt-bike champion Jim Scaysbrook's Gladesville shop on Thursday evenings. Most nights we'd pick a destination out in the country — usually a pub — and street-race out there, have a few beers and fang back. This was pretty irresponsible behaviour, I suppose, but by God it was fun. In the early hours of Friday it was common to see eight or ten of us hooning down the highway in tight convoy on our way home. These Gladesville meetings eventually led to the formation of a club called the Crash and Burners, which is still going today, although the original Crash and Burners are getting pretty toothless and hairless.

One of the Crash and Burners had a property at Tuena, down near Crookwell, and this became our weekend motorcross track. We'd camp in the old shearing shed and race all weekend, incurring some horrendous injuries because in our number there were some extremely good riders who forced the level of skill beyond the limits of some of us.

Ian wasn't a bad rider but he was braver than he was skilled, and he used to crash a fair bit. He always pushed the limits until he either beat them or suffered the consequences. I remember him dislocating his shoulder quite badly once. But he was mostly famous at the Crash and Burn weekends for having the most revolting food. He arrived one weekend in his builder's ute

> *full of junk. It was lunch time and he pulled this filthy Esky out from under his bike and a bunch of shovels and produced a string of sausages that had been floating around in the melted ice for a couple of days. He hung them over the rails of the ute to dry out, then he cooked them. It was disgusting. He used to bring a sailing mate from New Plymouth and they'd usually get warmed up along the way. One year they got lost and spent twenty-four hours at some country pub. When they arrived they looked like they'd been run over by a herd of buffalos. In the early days one of the highlights of the Crash and Burn weekends was the Saturday afternoon session at the Tuena pub. The trip back to the property was always a lot faster than the one in! Ian actually turned his bike over in the middle of the main street, trying to lead the pack out of town. We stopped going into town for a drink after that. We didn't drink any less, we just did it where there was less traffic.*
>
> Jim Scaysbrook, motorcycle champion

Since we'd pulled off the Globe Hotel deal, Gary Bogard of Home Units Australia had become quite a good friend. He'd sailed with me a few times and had started to develop an interest in yachting, so much so in fact that he was talking about getting a boat for the Admiral's Cup in 1973. (He did, in fact, gain selection in the Australian team and my friend Dick Hammond was his navigator.) I liked Gary's up-front attitude to business and I respected the way he quickly summed up the good and the bad aspects of a deal. In turn, he seemed to think I was a builder of at least average intelligence, and a bloke who had a bit of a feel for the market.

Knowing that I was soon to sail for New Zealand, Gary invited Judy and me to a Sunday lunch by the pool at his beautiful home in Attunga Street, Woollahra. Over a bottle of wine he suggested that we further our joint-venture involvement by setting up a plan to purchase 500 terrace houses in the inner city. I was somewhat taken aback by the enormity of this undertaking, but Gary and his partner Sid King usually talked in telephone numbers, and so far we'd done okay.

'Consider this,' said Gary. 'If we buy 500 houses and they are all washing their faces [paying their way], then if they go up in value by $1000 we make $500,000. If they go up by $5,000, we make $2.5 million.' The scheme made perfectly good sense because I knew there were 500 houses around the edge of the city that we could pick up cheaply, and, provided we renovated cheaply and effectively, we could have them let and covering their borrowings in a matter of months. Two factors were crucial: access to virtually unlimited borrowings at sensible interest rates, and a renovation 'machine' that could take advantage of the economies of scale without losing the 'trendy' feel of our existing rented terraces. If Home Units Australia could get the money, I could make the places attractive and secure top rents.

A few days after the lunch I sent Gary a detailed proposal for 'Joint Venture 500':

> Of course it is easy to incorrectly consider that the situation is static and lose sight of the fact that every year there are more kids going to Uni, more getting out and flatting, getting married and so on, and that the terrace house is highly fashionable as well as being a diminishing commodity. As well, the terrace offers far more than an equivalent priced flat: it has more character; some garden versus nil; more privacy; a terrace can have a touch of fashionable dereliction, whereas in a flat this would not be acceptable . . .
>
> . . . In this early stage it would be better to try to limit the renovation costs to the absolute minimum. As I see it, the cheaper $30-40 market is the easiest to let, and where there is the least currently available. The kids today would rather have an outside loo painted bright yellow with a big daisy on the cistern and a rent of $35, than a tiled bathroom with a rent of $45 . . . I imagine we can retain a really top designer who will help us arrive at a schedule of standard

> finishes . . . For instance, ochre, red, green or similar colour to exterior. Stencilled street number and window box with geraniums or similar . . . Kitchen: one old roundline style refrigerator painted bright orange or similar with large graphic on the door . . . painted concrete floor . . . set living room modular furniture, bean bags, wall hangings etc . . .
>
> If we can assemble a team that will . . . march into a house, apply a selection of those items . . . and create a comfortable, fashionable, cheap house for $1,000-3,000, we can start to churn out these houses like the proverbial sausages.

If you were around inner Sydney in the early 1970s, I am sure you will recall the Joint Venture 500 terraces. We did churn them out like sausages and they became the standard low cost rental accommodation for the new type of city-dweller.

Gary Bogard responded enthusiastically to the proposal and we bought a shelf company called Tierra Del Fuego Pty Ltd as the vehicle for the joint venture in which Gary, Sid King and I had a third share each. I phoned estate agent Spencer Simmons and told him to start finding houses, and then I sailed for New Zealand.

My crew for the voyage consisted of Greg Walch, a Tasmanian sailor I'd first met on Lord Howe Island, and Robbie Venables, a rigger from Careening Cove who reckoned he could navigate. (He couldn't, as it turned out, but we were well clear of the Heads when I found that out!). This Tasman crossing was a landmark voyage for me on *Maris* and I spent a lot of time preparing for it. Although I thought I'd have navigational backup from Robbie, I got Jack Earl to give me a last-minute drill on celestial navigation. Then, in late November we set sail in favourable conditions and I decided to sail directly for Cape Maria Van Diemen at the northern tip of New Zealand, rather than head for Lord Howe. Four days out the wind went to the southwest and we found the massive Southern Ocean swells. What a thrill to surf down these long green faces! Each of us would wait for the watch to change so we could take our turn at the helm and try to surf faster than the last guy.

As we closed on the New Zealand coast I began to get navigator's nerves. Jack had taught me well and I thought I'd plotted my celestial course well enough, but he'd given me a bottle of rum and a letter before we left, and now I went to the chart table while I read the note. It was basically a letter of encouragement, but it had some important advice about landfall. If there was no landfall as calculated, Jack said at sunset I should slow the boat down, have an easy night and keep a good lookout. The next morning if land was still not in sight, we should have a good, hearty breakfast; then we'd have plenty of daylight to run down our landfall.

Jack was dead right. I slowed down and at eight the following morning, we had the Three Kings Islands in sight. It was an exhilarating feeling. We had crossed some 1,600 kilometres of ocean guided by the stars, and we were within three or four kilometres of my calculated position. We sailed fast now, and by mid-afternoon we were close to Cape Reinga, looking shoreward to picture-postcard country of rugged green hills. By the next morning we were off the tiny entrance to Whangaroa Harbour, and in high spirits we sailed into this beautiful little bay and tied up at the main wharf.

We were soon joined by Jim Bratty, the local policeman, a big, jovial fellow who arranged our customs clearance. When the paperwork was done he said: 'Have you got wet clothes?' We had. 'Stick 'em in a sail bag and bring 'em back to the wife.' We did as we were told and Mrs Bratty kindly washed our deck gear while Jim showed us around his courthouse, which was attached to the house. He sat at the magistrate's bench and told me to sit in the dock. He glowered down at me for a long minute, before breaking into laughter. Still, I got a clear indication of what it would be like to be on the wrong side of the law in Whangaroa.

At the Marlin Hotel opposite the wharf the local fishermen bought us round after round and explained where all the good anchorages were along the coast. We slowly worked our way towards the Bay of Islands and five days later sailed into the fishing village of Russell. *Maris* was well known here because Jack and his son Mick had put in for provisions and rest in 1962 on their way across the Pacific, when they set a new trans-Tasman record for a boat under fifty feet. The record of seven days and one hour stood for several years. And here at Russell there was a letter waiting for me from Jack. It was a beautiful letter and it set the tone for so many he would send me when I was sailing. He looked at the weather maps and the charts, and he knew exactly what I was experiencing on every voyage. It was almost like having him sitting on my shoulder as I crossed oceans, and I can't tell you what a comfort that has been at times:

> Ahoy Ian — Every hour of the days since you and your two shipmates waved as you put out of Mosman Bay aboard that gallant yawl I have had pictures of you roaring on to New Zealand — by the weather maps, a wild trip . . . Reckon you will be about 250 miles off Cape Maria Van Diemen now, a dark blowy night with plenty of water over your little ship under reduced canvas — reefed main and No 2 jib or mizzen and No 2 jib. Here is to a good landfall, old pal. When you get this note you will have made it so we will raise our glasses to you for a mighty effort.

I sent off a note in reply: 'Cape Reinga exactly where you said it would be. We felt like Three Kings!'

That weekend, on Waitangi Day, we raced with the Waitangi Yacht Club and finished a close second to the commodore in his 37-footer. Then we pushed south to Whangerei and eventually Auckland. Robbie and Greg were both sailing in that year's Sydney to Hobart and I wanted to spend Christmas with the girls, so we secured *Maris* and flew back to Sydney.

It was a good family Christmas, but I felt a great sense of relief to rejoin *Maris* on 2 January 1973. It wasn't just my concern over the safety of the boat. Despite my best efforts to avoid it, my passion for sailing and the adventurous life was beginning to drive a wedge between Judy and me. We both tried to paper over it, for the sake of Sally and Pip as much as anything, and there were still plenty of good times enjoyed together. But increasingly, we were going our own separate ways.

I had to find a shipmate, so I put a note on the board at the yacht club and a young farmer came down from Hawkes Bay and offered to crew. Will Finlay was a steady sort of Kiwi. I signed him on and we set sail from Westhaven, cruising at a leisurely pace to Whangaroa as I taught him the ropes. I had no self-steering on *Maris* at this stage, so it was two hours on watch, two hours off, all the way back across the Tasman. It was a long, slow crossing, but the boat handled beautifully. I wrote in my log as we neared Sydney: 'Great miles, little *Maris*, just starting to realise how wonderful you are.'

I hadn't been home five minutes when my old school mate from TAS, Ralph Featherstone, arrived from his cattle property near Grafton. Ralph had a gift for us, a pup sired by a blue cattle dog from a pure-bred dingo bitch. Smokey was a real beauty, smokey-blue in colour with the dingo's white-tipped tail, but Alf loathed him from the start. Smokey thus presented a problem.

We were about to move into high gear again at Ida Pender Place and I had been negotiating with a number of rent-controlled tenants there. Most had agreed to vacate for a cash settlement, some with low-rent alternative accommodation provided. But I had a problem with a bloke called Bert Fleming who, with his blind wife Norma, occupied the basement of a house right where we wanted to build the kitchen for a restaurant. I went to see the Flemings with Smokey in tow and we struck a deal.

Later I outlined the agreement in a letter, explaining that 'my company would provide them with [a double-flat in a block across the street] for the duration of both their lives. Mr Fleming also agreed to act as caretaker and watchman over the complex for a period of ten years and for this purpose I supplied him with a dog.' Two problems solved and Bert and Norma were as happy as Larry.

Tierra Del Fuego was on the march. We had properties

everywhere, even a couple in Melbourne, in various stages of renovation. Two days a week I would go out cruising in Judy's air-conditioned Mercedes, looking for purchases. The real estate agents would point out available rows of terraces and sometimes I would simply circle the block before saying yes or no. I had come to know the inner city so well that I could judge the condition of a terrace without going inside. Mind you, this was not always a good idea.

One day I was having lunch with one of the Crash and Burners, an advertising man named Mike Robinson, at an East Sydney Lebanese bar where Tre Scalini Restaurant is today. It had been pouring with rain for more than a week, but suddenly above the din of the rain we heard a horrendous roar and a crash. We ran out onto the street and I saw that three of my terrace houses had fallen down in a pile of rubble. It ruined the lunch, I can tell you. In fact we lost five terraces in about a month.

Since my initial successes in the restaurant trade, and a 1971 venture with a couple of mates called Sid Preston and Robyn Fiorelli (we opened a restaurant in Willoughby called Tummies), I had been keen to tap back into the cash flow of restaurants. We had put a restaurant, bar and disco into Squizzy Taylor Square, and I looked forward to completing the state-of-the-art restaurant we were putting in as the $250,000 centrepiece of Ida Pender Place. Managed by Sid Preston, Ida's was going to be a real statement about our approach to inner-city development, showcasing the restoration work we'd done, alongside totally modern concepts, like our lighting effects and pop art panels by Hanna-Barbera, the American animators who had set up a Sydney studio.

All of this was costing a fortune, but finding funds was not a problem. Almost every day I fielded phone calls from finance companies who would ask, 'What have you got coming up? Put up a deal to us.' Some went even further, offering 110 per cent mortgages which meant you could call this a bonus and have some cash for the next deal. We stuck with the lenders we knew, but by the end of 1973 even they seemed anxious to sell us any amounts of money. It was property boom time in Sydney. The opening of the Opera House in October signalled for many the birth of a new world city, but as so often happens at the peak of a boom, the storm clouds were already gathering. Prices were escalating out of control, but so were interest rates. It seemed only a matter of time before the storm would break.

I sailed on *Ragamuffin* again that year. One day after a race Syd Fischer asked me how business was going. I told him I was just a little worried about the economy. 'Mate, we're in for the biggest fucking gale we've ever seen. Better batten down the hatches.'

> *I was getting very worried. We were paying too much for houses. One we bought in Liverpool Street cost $90,000 and was worth about $25,000. I felt Ian was a bit too over-enthusiastic about it all. I represented the conservative view and he was being driven more and more into the HUA camp.*
>
> Gordon Stewart

By now my joint ventures were extremely complex, with cross-financing and refinancing schemes covering the entire grid. Home Units Australia had a piece of Squizzy, a piece of Ida and a piece of another development we'd started in Surry Hills, and King, Bogard and I had equal shares in Tierra Del Fuego. Our companies had massive borrowings, but the difference between mine and theirs was that most of mine were personally guaranteed. In early 1974 I sat down and analysed the situation with Lee Norman, the general manager at HUA and an old rugby mate of mine.

Tierra had 389 terrace houses worth an average $15,600 each. The trouble was the market had overheated and people had stopped buying. The agents were saying they couldn't sell terraces in Paddington, so what chance did we have of offloading half of Woolloomooloo, Darlinghurst, Surry Hills and Redfern? Many of the houses were attracting good rents, but these nowhere near covered our outgoings during a period of heavy development. To keep the whole show going, we had taken to borrowing to pay our interest.

Lee Norman looked up from the pages of projections and said: 'I reckon we're fucked for bricks here.'

In June we opened Ida's Restaurant to immediate critical acclaim and full houses. Sid Preston knew his way around a restaurant and had ensured that food and service ran smoothly from opening night, but there was more to Ida's than food. It was open and airy and looked out over the city. It had a real Sydney feel about it, and people like to know what city they're in.

A few days after the opening, restaurant critic and bon vivant Leo Schofield gave Ida's a resounding thumbs-up in his *Sunday Telegraph* column:

> I am willing to concede that the beautiful sunny day, the open terrace, the pop decor . . . and the contrast between the mutilated city with its monstrously ugly developments and this sensitively organised and stylishly executed ambience had something to do with my having such a pleasant lunch, but the food isn't half bad.

Leo went on to document a meal which sounded delicious and, by the prices, a long time ago. 'Mrs Schofield opted for the Queensland Mud Crab Bisque ($1.40) which was very good indeed. Served with Garlic Damper which I thought a bit camp.' Yes, it was a long time ago.

The excitement of the positive reception Ida's was receiving from Sydney's 'trendies' momentarily lifted our spirits, but every day there was increasing evidence that my property empire was in grave danger. Inflation, which had stood at a reasonable 4.5 per cent when Joint Venture 500 was conceived, was now a staggering 14.4 per cent and growing rapidly. Interest rates were at 17 per cent and who knew how high they would go? And my relationship with HUA began to change, although Gary Bogard remained friendly.

One night Judy and I accompanied Gary and his wife to a B.B. King concert. Gary was a big blues fan and I knew something was seriously wrong when he kept ducking out during the performance to make phone calls. Even when he was with us during and after the show, he was distracted and nervous. At one point when I had his ear, I asked: 'Mate, is there something wrong?' He waved me away with a hollow laugh.

The next morning, 9 July 1974, I was on site at Ida Pender Place when I got a message to call him. His voice was barely audible and before he started I knew I wasn't going to like what I heard.

'I'm sorry, mate. We're out of business.'

Left: Mum and Dad (Leslie and George) on their wedding day, 5 December 1936, at St Mary's Church, Farnham Royal.

Below: Jack and Kathleen Earl beside 'Myee' at Dangar Island, around 1928.

Above left: Wearing Dad's slouch hat during the war.

Above right: Mum, me and Ann at Gibsons Beach, Watsons Bay in Sydney Harbour, in 1941. From the start I was tantalised by the water.

Left: Me, Mum and Ann visiting the city in 1945. Cop the sailor suit!

Above: Dad, Mum, Ann and me at Gillingwood, 1946.

Right: Looking to the future in 1956, aged sixteen, in a portrait by Monte Luke.

Above: With my schoolmaster and friend Jim Graham (left) on the front field, TAS (The Armidale School), 1957.

Right: Building at Kirribilli with 'Siggie', my German bricklayer, in 1966. The terraces face onto Sydney Harbour.

Above: My first wife, Judy Allan, on our wedding day in 1966.
(MONTE LUKE)

Right: Building my first yacht, *Bimblegumbie*, in the backyard, 1967.

Right: Alf, my first blue cattle dog, standing guard outside the row of terraces IBK had just renovated in Cathedral Street, Woolloomooloo. All three terraces cost $20,000, but I sold them in my first deal with Home Units Australia for $80,000!

Above: Launching *Bimblegumbie* on the day Pip was born, 1969.

Below: (From left) Dad (obscured), baby Sal, Judy, my Grandma Bruce and me outside Superba Parade, Balmoral, in 1968.

Below: Squizzy Taylor Square, Crown and Liverpool Streets, East Sydney. From the street after the renovation in 1970.

VD 382

LIVERPOOL STREET
SQUIZZY TAYLOR SQUARE
CROWN STREET
ONE WAY
214
BRITISH RETAIL PROMOTIONS
PTY. LTD.

Above: Solo on board *Maris* in a stiff winter westerly off Sydney Heads.
(*ACE MARINE PHOTOGRAPHY*)

Above right: The wreck of *Maris*, Coffs Harbour, 1971. One of the worst days of my life. I'm emerging from the surf having attached the tow line to the boat.
(COFFS HARBOUR ADVOCATE)

Below right: A crowd gathered as *Maris* was slowly dragged up the beach.
(COFFS HARBOUR ADVOCATE)

Above: Sailing *Maris* in mid-Tasman in 1972. Note the bacon drying in the rigging.

Right: Californian shipmate Mark Fraser at Lahaina Yacht Club, Maui, 1975.

Right: *Maris* beside the remnants of the World War II dock at the deserted island of Palmyra in 1975.

Right: Hawaii bound, 1975. At the helm on *Maris*.

Left: With the Bilge Rat at Mosman Rowing Club in July 1977.

Below: With Carole Laird the night the Admiral's Cup team was announced in 1977 at the Cruising Yacht Club of Australia.

Above: The Admiral's Cup Team in Cowes, 1977. Far left rear is John Bertrand, far left kneeling is Tom Stephenson, sixth from left rear is Sir James Hardy, Syd Fischer is at the helm with Dick 'Sighty' Hammond directly in front. I'm sitting on the rail at the far right.

Right: On the quay in Papeete, Tahiti, in 1979 when Judy (née Humbley) flew up to meet me.

Above left: Celebrating my fortieth birthday at the Crash and Burn motorcycle event in the Woolshed at Tuena. The iced cake was cow dung – when I picked it up all the cow dung beetles ran across the table.

Above right: I made friends with Ua Po when we were nursing our burns in 1975 in Pago Pago. Pictured here with him on board *Maris* in 1980.

Right: The fish kept getting bigger – a good catch on board *Maris* south of the equator homeward bound in 1980.

To Ian and Maris —

All our love and luck old shipmate's
for another tremendous voyage together.
To your wild capable hearts and
love of adventure, go all our
blessings and admiration.
A happy Christmas Ian and a
fabulous 1980.

Kath and Jack.
Maris Tomo. Tiare. Loni.
and Matt.

Above: A beaut note from Jack and Kath Earl that was waiting for me at the post office in Lord Howe when I arrived on my way back from the 1980 Trans-Pacific.

Above: With my old Dad, George Kiernan, after my return from Pacific adventuring in 1980. He died the following year.

Right: My daughters, (from left) Pip and Sally, at their home at Mosman, 1982. (*SANDY HAMMOND*)

CHAPTER SEVEN

SAILING AWAY

It took a while for the ramifications of Gary Bogard's sombre announcement to sink in. Exactly who was out of business? I was a streetfighter, I wasn't about to go down without a blue.

Slowly I began to get the full picture. Colonial Sugar Refineries (CSR) had taken a 50 per cent stake in Home Units Australia through their building products division in July 1973, making them equal partners with King and Bogard. At the same time the personal guarantees of the HUA duo had been replaced by those of J.J. Finance and Norper Investments. This financial restructuring meant nothing to me at the time, other than the possibility of more finance being available for expansion. It was to mean a lot more down the track.

A year later, CSR's exposure was across a broad front which included Joint Venture 500 and the big Pallisades development in Kings Cross. They had been worried about the impending recession for months, and as construction and development companies began to collapse, they passed their concern down the line to King and Bogard, who passed it on to me. Not that I'd been unaware of the economic climate. I'd been horrified by the collapse of Mainline, the construction company run by the brother of my friend Harold Baker, the solicitor who'd given me such sound early advice about gearing. Unfortunately Harold's brother had been like me, and not heeded the advice that in the property game you never invest long and borrow short.

Now CSR had pulled the plug on HUA, preferring to cut their losses at around $5 million rather than slowly bleed to death. The finance companies panicked and called in the loans. Overnight HUA was sunk and Gordon Stewart and I appeared to be dead in the water. I called a crisis meeting with Gordon and was impressed with the way he calmly and methodically analysed the problems we faced. Although he wasn't involved in the Tierra Del Fuego joint venture, he was involved with HUA in Ida Pender, Squizzy Taylor and several other smaller property deals. He didn't have the solutions but he was prepared to look for them, rather than cut and run. And one of the first things we realised was that if we were to save anything at all out of the wreckage, we had to complete our existing development projects.

The week that HUA went down, IBK Constructions received no construction draw and couldn't pay the wages at Ida Pender Place. We had no alternative but to temporarily stand down our staff of eighty and another forty sub-contractors while we hurriedly sought an agreement with the financiers to turn the tap back on. Still, with Gordon holding firm, I clung to the belief that we might just be able to talk the lenders into going into a holding pattern until we could unload some of the properties.

There were two things wrong with this plan. The first was that with interest rates criminally high, the people who would normally be in the market for our terraces could not afford to buy them, and the investors weren't touching their wallets until the fog lifted from the economy, and with the Whitlam government in deep trouble over its attempts to raise Arab loans, who knew when that would be? The second thing wrong with it was that the sharks of the finance world had begun to circle, ready to snap at the assets of the property businesses hitting the wall every day.

On 25 July, just a fortnight after the HUA collapse, the finance company FNCB and property giant Hookers announced a 'restructuring' of HUA. In this 'rescue arrangement' CSR was said to have reduced its shareholding to 25 per cent, while Hookers had purchased 25 per cent and FNCB 50 per cent. Former directors Bogard and King had sold their shares and retired. (They both relocated to Monaco soon after.)

FNCB-Waltons Finance Ltd, a joint venture between the First National City Bank of New York and the Australian retail and investment outfit, was thus the major shareholder in both Ida Pender Place and Squizzy Taylor Square. Just a couple of weeks before the HUA collapse, FNCB-Waltons had appointed a receiver when Queensland developer J.R. Roberts defaulted on loans totalling $9.4 million. The loans on Ida and Squizzy were about half this, but both developments were good financial prospects for the future and the sharks had their evil eyes on them. Despite our pleas for a reprieve to enable us to auction some assets, FNCB-Waltons appointed a receiver-manager, a fellow called John Garrity.

I began to see what was going to happen. The sharks were snapping. In my view it was madness to sell the terraces at fire-sale prices and push the market down still further, yet that seemed to be what the finance companies were urging us to do. At the same time, however, they were loath to liquidate and put the whole lot on the auction block, for fear of making public their exposure and panicking their own shareholders. The whole thing was like a game of musical chairs. The music had stopped and everyone was trying to grab a chair. Or a life raft.

The financial crisis had placed tremendous pressure on my marriage, a union that had already suffered several stress fractures. Since her initial foray into gearing several years before, Judy had become more involved in the workings of IBK. She was a director of many of the companies in the group and had her own ideas about what we should do. These most certainly did not include my preferred option, which was to sell virtually everything we owned, including the big house at Moruben Road, the cars, the bikes . . . Sell everything except the boat, bring the lifestyle down a few notches, regroup and try to trade out of this mess.

Judy felt we should sit tight, negotiating with the financiers, selling what we could when the price was acceptable. I felt it was becoming impossible to negotiate with these bastards who seemed intent on ripping apart everything I'd worked so hard to create. Judy and I never fought over this difference of opinion, perhaps because for some time now she had felt some responsibility for the direction the companies had taken. We kept the peace at home because the last thing either of us needed was to open up another battle front.

I gave vent to my real emotions alone. In my anger and frustration I was expressing a feeling I could not shake, a knowledge deep within me that I was not going to survive this financial and personal dismemberment if I sat around watching it happen. It may have been unfair on some and seemed unreasonable to others, but I knew what I had to do.

Months before the full extent of our financial problems became apparent, I'd begun to plan a voyage across the Pacific and back. I figured I needed six months to sail through the island groups to San Francisco, turn around and come back. Just when I could fit this into my working schedule I hadn't worked out, but as the months went by it became a regular conversation item when I was at the Earl compound at Mosman Bay. Young Benny Hawke, who'd started a career in journalism, was raring to go. His mother Maris, who'd been wild enough in her day, had seen it all before and had cruised on her namesake herself. Boats had always been a part of her life and she had no fears about Benny coming with me.

> *Mum wasn't all that impressed. I had a good job in journalism and I was just tossing it in to go sailing. But it was in the blood. She had great faith in the boat. I don't know how much faith she had in Ian.*
>
> Ben Hawke, television producer

I had another crewman in the wings in Micky Thompson, a mate from Lord Howe Island. Micky knew I was planning the voyage and came to see me on one of his visits to Sydney, dressed in his Sunday best. I knew what he was going to ask, so I beat him to the punch. 'Micky, do you want to crew on *Maris* across the Pacific?' He was ecstatic.

Now, with pieces of the IBK empire splattered all over Sydney, I started to plan for a September escape. I sold what viable businesses I could. The restaurants, Tummies and Ida's, both went and I was able to get out of the Mendoza joint ventures and liquidate what properties I owned by myself. I got out as much cash as I could and gave most of it to our family friend, Bob Lane, someone I trusted implicitly. I said: 'This is the fighting fund, Bob. Hang on to as much as you can for when I get back, but don't let the family starve.' Then I signed the house and management of the company over to Judy and started provisioning the *Maris.*

> *I got left holding the baby, and I was extremely disappointed about that. Bogard and King buzzed off and Ian took off on his yacht! I was still very attached to Ian and I don't think I felt any real resentment about his leaving, because there was nothing much he could do to save the situation. But I was the one left on the mat and I was disappointed. It took me years of negotiation to settle the debt and in doing so I lost most of my own property. Maybe I've never told Ian this, but I was pretty disenchanted at the time. I suppose the friendship won through though, because on several occasions I settled financial claims on the condition that he be released from his personal guarantees as well as me from mine. The other upsetting thing was that his marriage was clearly on the rocks because of the pressures of the collapse. I liked Judy, I thought they were good for each other, so that upset me.*
>
> Gordon Stewart

I don't expect too many people who read this to understand why I sailed out of Sydney Harbour, leaving a huge trail of debt behind. My friends understood — even Judy understood that we all react differently in a crisis — and that was enough. My rationale was that this wasn't a bad tactic. The financiers didn't understand my terraces, my formula, they didn't understand me. They couldn't fix what they didn't understand and they wouldn't let me do it, so why not let them stew in their own juice for a while? I felt tremendous responsibility to my family and to my trades people, so I made sure that they were taken care of before I left, but I can honestly say that as I made ready to sail I felt not one iota of responsibility to the lenders who'd been so quick to throw money at me in the good times, and so quick to destroy me in order to get it back when things went bad.

I simply wasn't into the public opening of veins. Perhaps if I had stayed and fought at that point I would have been a millionaire again in five years' time. Perhaps a lot of things, but I doubt very much that I would have had such a rich and rewarding life.

The plan was for the Bilge Rat to sail with us to Lord Howe Island, then fly back home after we'd picked up Micky. In the week before we left the Rat and I worked long into the nights preparing the boat and finishing off the first of the Jack Earl-designed 'little pig' dinghies that I later made commercially. As well, I'd packed a second No 1 jib and at last installed self-steering. We were well equipped for the long ocean crossing.

There was a small crowd on the marina to see us off at eight on a chilly September evening. The girls were there, not quite old enough to grasp that they wouldn't see Daddy for a long, long time. Maris took me out of earshot of Benny, fixed me with her steady gaze and said: 'You know that if you drown him I shall haunt you forever.' We were very close but I knew she meant it. She was charging me with the awesome responsibility of looking after this wild twenty-year-old. I told her straight that I accepted that. Kath and Jack Earl stood in the garden of the compound overlooking Mosman Bay as we slipped past. Jack blew a dirge on his Darnley Island conch shell as Kath waved a big Red Ensign. On board *Maris,* the Rat, Benny and I responded by dancing a jig on the deck and whooping and hollering at the night.

We were away now, by God! Let the sharks kiss my arse! And before we had cleared the Heads the waterfront was alive with rumours that Bicky Kiernan had sailed away from trouble with a keel of solid gold.

As we sailed out into the Tasman, bound for Lord Howe Island, we hit one of the most savage gales I had yet encountered, with bitterly cold winds whipping up a huge sea. In the end we hove to, sharing the freezing watches and listening to the broadcast of Parliamentary proceedings where Gough Whitlam was taking a hammering over the economy. The Bilge Rat said: 'That poor bastard's gettin' a worse belting than we are!'

As always, we sailed with Jack Earl on spirit watch with us. A couple of days after we arrived at Lord Howe, a Mosman yachtie followed us into the lagoon with a personal delivery letter for Ben. Jack had written to his grandson at the height of the gale:

> Ahoy Ben — We are all concerned over the violent weather you have experienced, as I know you would be out too wide . . . to go for Broken Bay. The big low in the central Tasman . . . must have given you full gale conditions, bitterly cold with driving rain and a bastard of a sea . . . We had gales here . . . the coldest for fifty years, snow in Sydney and bad from Tasmania to Queensland . . . Guess Ian had *Maris* down to a spitfire, extra lashings on the dinghy and trailing lines. Hope the wing from your self-steering was below, and hatch slides in . . .

There was nothing the old man didn't know about *Maris*, and he sailed every mile with us.

I celebrated my thirty-fourth birthday on Lord Howe Island, then the Rat flew home and Micky Thompson joined our crew for the voyage to New Zealand. The crossing was good, with clear days and following winds. We made the landfall with ease and we cruised down the familiar coast, renewing acquaintances and raising a little hell. At the Marlin Hotel in Whangaroa Harbour we held the pool table for four days. Micky's brother Nat Thompson was on a boat called *Utiekah III*, a rundown old Tasmanian ketch skippered by a hippie we dubbed 'Gary Glitter'. We partied and sailed with the local characters for a week or two, then sailed down to Whangarei and Auckland, before striking out through the Kermadec Islands to Raratonga in the Cook Islands.

In Raratonga we hired motorbikes and toured around the island, stopping at villages for feasts, terrorising the Polynesian dance halls . . . I'd been away from the fall of Joint Venture 500 for almost two months now, and it was beginning to feel like it had never happened.

One morning I came up on deck and there was a very formal Cook Islander waiting to see me. He wore a shirt and tie and real leather shoes. 'You are the master of this ship?' he asked. I admitted as much. He introduced himself as a cabinet minister in the Cook Islands government and asked, very politely, if I would be good enough to come and meet with the Prime Minister, Sir Albert Henry. Figuring that we were either about to be deported or given the keys to the city, I slipped into a fairly clean tee shirt and a pair of cut-off jeans and went off to meet the most important man on the island.

Sir Albert was genial enough, but he didn't mince words. He needed to charter a boat and *Maris* looked the most appropriate for what he had in mind. I did little more than nod, and we were in the charter business.

It was election time in the Cook Islands and Sir Albert was facing a real challenge from Opposition leader Tom Davis. Davis was, by all accounts, a nice bloke while Sir Albert was a bit of an old villain, but business is business. And Davis already had a boat, *Moana*, which was owned by his son. The job was to deliver ballot boxes, scrutineers and gifts to the elders on the outer islands. 'How much?' asked Sir Albert. Never having chartered before, I had to think on my feet. I said 1,000 American dollars would cover it, plus food and fuel. We had a deal.

The cargo started arriving on the wharf and we loaded *Maris*.

Eight bags of mail somehow got into the deal, then people brought parcels for cousins and soon we were loaded to the water line and just able to squeeze through the cabin to reach the chart table. The scrutineers arrived. One had two cases of beer which he sat on in the cockpit while he began the task of emptying them; the other was a man-mountain who formed a human barrier on the other side of the cockpit, making it very difficult to work the boat. Just as we were about to cast off, my fare-paying passenger arrived. I'd forgotten about him. I slipped his bank note into my pocket to divvy up with my shipmates later on in the approved split of one part each, one for the boat.

As soon as we were out of the harbour I realised that we should have delayed our departure. It was squally with a nasty sea building and it was heavy going all the way. Every wave that slammed into us threatened to send the ballot boxes, food parcels and mail bags flying around the cabin, testing the lashings that restrained them. The power of the sea made no impression at all on the two scrutineers. We arrived at the island of Mitiaro, which is just an atoll really, and the villagers rowed out a long boat for us to unload our cargo. We drew straws to see who stayed on board and Micky lost.

The hospitality of the chief and the villagers was extraordinary and they wanted us to stay and party on into the night, but I was determined to leave in daylight. The incoming tide had pumped the rising swell even higher and when we got back to the *Maris* Micky was in quite a state. He had our biggest anchor down and it was barely holding us. We got out of there as fast as we could and made for blue water with the last light of the day. We sailed all night while the scrutineer slept off his beers, and in the morning we were off Mauke. The sea was even rougher here and we decided against going ashore. Micky wasn't too happy about it as we unloaded the remaining cargo onto a long boat and lowered the sleeping form of the scrutineer aboard, along with his four remaining warm beers. Forty-four for the voyage may not have been a Pacific Islands record, but it wasn't a bad effort.

The hurricane season was approaching so we made our move north and east, away from the chain of thirty-two islands. My plan was to sail to Bora Bora, and then to Papeete, but the weather was so bad we went straight on to Papeete. The conventional wisdom amongst Pacific boaties back then was that once you got east of Bora Bora you were out of the hurricane belt. Not so, as it turns out, but I had no reason to doubt it at that time.

In Papeete's beautiful harbour, protected by the outlying coral reef, you anchored 'Tahiti style', which was stern on to Quay Galiano, the main waterfront boulevarde. Here there were boats from all over the world and sailors of all kinds, from lean and tanned hippy cruisers with the bums out of their jeans to the legends of ocean sailing. People like the 'papa' of single-handed sailing, Bernard Moitessier, who was there on his yacht *Joshua*, and Mark Darnois, who had a huge French schooner called *Maylis.* Darnois was a sun-blackened Frenchman with long white hair and only one leg, the result of being shot down as a fighter pilot during the war. His story was that the doctors amputated the wrong leg, then had to fight to save the damaged one.

Papeete was the biggest city in Polynesia — the only city, really — and the jumping off point for Bora Bora, Moorea and the Marquesas. It was a place where crews split up or found each other, romance blossomed in the creaking cabins of ancient boats and people floated by in a haze of dope and booze. Some had succumbed to tropical paralysis and let their boats decay around them, but in the main I found it a lusty, spirited place, alive with adventure and excitement, and the memories of the creative souls whose art had flourished there — Gauguin, Conrad, Stephenson . . .

Two letters from Jack were waiting for me. Again I marvelled at how much detail he seemed to know of our voyage just completed. 'It's hard to believe you are right out there in the centre of French Oceania, bashing away to windward in those legendary seas with rain wiping out your landfalls. Poring over the chart and checking your DR [dead reckoning]. Taking hurried shots of a hide-and-seek sun with Ben and Mick running the ship off to steady her, no assistance at all from the French weather reports . . .'

Jack was spot-on with one exception. As we beat windward to Papeete, Micky lay on his bunk in agony, the victim of a vicious hernia. It was going to be months before he was strong enough to work the boat, so reluctantly we put him on a plane for home.

Maris was well known in Papeete, thanks to Jack's Pacific journeys, and we soon clocked in to his circle of French and Tahitian friends. We became particularly close to a guy called Rodo Williams, a man with classic Polynesian features made even more striking by the fact that he was blond. Rodo had a story about everything, and his story about being blond was that his hair had gone from dark to light overnight when he spent more than twelve hours swimming with a nine-year-old girl on his back when a boat capsized. Rodo had two sons, both called Michel. One was blond and the other dark, so there was never any confusion. There was also never any doubt in my mind that his blond colouring came from the genes, rather than an all-night swim. He was married to a Cook Islander named Queenie and they lived in the traditional Polynesian style outside of Papeete.

Rodo came from a Tuamotu seafaring family originally from Katiu, a small atoll about 500 kilometres east of Tahiti. At some point an American seaman called Williams had married into the family and they had adopted his name.

Just a year after I met him Rodo was the pilot on the historic voyage of *Hokule'a*, the double-canoe that sailed 10,000 kilometres from Honolulu to Papeete and back again to prove once and for all that Polynesian migration was a planned move, rather than a fluke of wind and tide. Rodo's specific task was to find a sea road through the treacherous waters of the Tuomotu chain. According to Ben Finney's 1979 account of the voyage, *'Hokule'a: The Way to Tahiti'*, Rodo knew his home waters but not the celestial navigation he had supposedly learnt at his grandfather's knee in the 1930s. At one point he produced a map that had been brought along by the cameraman and tried to identify a landmark. Finney claims he plucked the map out of Rodo's hands and threw it to the wind, since maps were banned on *Hokule'a*. 'No problem,' said Rodo.

If Rodo knew his way around a boat, he wasn't so good in his tiny De Cheveux, which he called his 'cowboy car'. He'd drive around the narrow back streets on the wrong side of the road, deep in loud and animated conversation with us in the back while he ran over chooks and mounted the footpath. 'I no kill you,' he would laughingly reassure us after each near-fatality.

Rodo was the main man around the waterfront. We'd pull into a service station for petrol and Rodo would demand cold beers for his friends. The service station attendant would hurry around the back and return with a few bottles. He'd knock on some back-alley door and demand dinner, and we'd find ourselves enjoying a meal in a stranger's dining room. This rudeness worried me at first, but I discovered that these homes actually operated as unofficial restaurants, providing cheap meals for the waterfront population.

Rodo ran a fishing charter business and was justly famous for having dragged a 410-kilo marlin into the lagoon so that he could play the fish in front of an appreciative audience on the shore. We spent quite a bit of time at the Williams family compound, a plantation about twenty kilometres from town that stretched from a long line of beachfront all the way to the foot of the mountains. The *grandpère*, who was also blond, lived in an old house way up the back, where, as patriarch, he put on luaus for us. We would feast on roast pig and tell stories for hours, but the family lived by the sun, and soon after dark it was time for bed. Everyone slept together in the old house, although the daughters slept with their mother, to keep us at bay, I suppose.

> *Another of Jack's mates was a TV producer who worked around the Pacific a bit, and he'd been hired by a South African rum company to shoot a television commercial in Tahiti. He saw* Maris *in the harbour and came aboard for a few drinks. He'd already hired a French boat, but we talked him out of that and we got the gig. We sailed over to Moorea and Ian and I were the*

models. They flew in two Australian models to be our girlfriends, but because I was so young, my role was more that of the boat-boy, pulling up spinnakers while Ian sat in the cockpit with the spunks. I couldn't understand why they had to use these Australian girls when the place was full of beautiful Tahitians, and then the penny dropped. This was for South Africa, and white boys frolicking with coloured girls would not be acceptable.

Ben Hawke

We also became friendly with the people at the Tahiti Yacht Club and at their invitation we entered the Round-Tahiti race just before Christmas. This was more of a party than a race, and after a day and a night of sailing we anchored in a little bay at the bottom of the island called the Presque-Ile and partied from boat to boat. That night I met a very loud and very drunk American who insisted that I was a 'crazy Aussie bastard'. I figured that anyone who was such an astute judge of character could not be all bad, and Mark Fraser soon became a friend.

I was missing the girls badly and decided to fly home to be with them at Christmas, and to see how the sharks were devouring my businesses. We had plenty of friends around the waterfront now, and I didn't feel bad about leaving Benny alone for the festive season. I knew he wouldn't be alone for more than five minutes anyway!

When I got back early in January I jumped aboard *Maris* and was alarmed to see the cockpit a shambles. Down below Benny was on his bunk in a bad way. His leg was in a rough plaster cast and he was in a lot of pain. In between gasps he told me what had happened. Heading out to a party on New Year's Eve he'd been bowled over by a car and had broken his leg. The driver had got him to hospital where the doctors had slapped on a cast without properly examining the break. It was obvious that he had to go home for proper treatment.

I had been thoroughly enjoying myself as the master of the Maris. *In the afternoons I'd get dressed up in Ian's best clothes and have people in for drinks or I'd go and party on other boats. On New Year's Eve I went alone to a little Chinese cafe on the waterfront to line my stomach before a big night of partying. I'd only had a few beers with the meal, but I stepped onto the road and momentarily forgot which side the traffic was coming from. The guy who bowled me was a Tuomotan who probably wasn't used to traffic. He was very upset and whisked me off to hospital. After they'd treated the leg this guy wanted me to go out drinking with him, but I declined the offer and retired to the boat in pain. I was still lying there when Ian got back.*

Ben Hawke

A few days after putting Ben on a plane I heard from Jack. 'His X-ray brought bad luck and prompt action,' he wrote. 'The upper break was right across and the ankle showed a big chip of bone that had to be pinned back as it was completely separated. They operated within half an hour of taking the X-ray. We saw him tonight but he was too heavily sedated to talk much. The plaster could be on for a month or two, so it looks like you've lost another shipmate.'

Indeed I had. It was just me and *Maris* in Papeete with a long way to sail, and I didn't feel ready to undertake a single-handed ocean crossing. I began searching the waterfront for suitable shipmates.

CHAPTER EIGHT

A TROPICAL MALAISE

For a while there it looked like my new shipmate might be a girl called Leilani. She was an exotic, free-spirited Micronesian-American, looking for a ride to Hawaii, but she had no real experience of crewing and I could see that she might become a liability. We spent some time together on short sails out of Papeete, had a brief but energetic affair, and came to the mutual decision that for her to sail with me to Hawaii was not what either of us wanted.

Leilani and I remained friends, however. On one occasion I called in to Papeete and found her suffering from a bad tropical fever, semi-delirious and in a cold sweat. I stayed with her throughout the night, sponging her down, and the next morning got her to a doctor. Leilani continued wandering the Pacific and I kept tabs on her through her family for some years.

The waterfront was full of people eager to crew. The Tahitian government had instituted a bond system so that the state didn't have to cover the cost of off-loaded crews, but despite the bond that skippers had to post for each crew member they brought into port, the game of musical boats was still played and there were plenty of sailors hitching rides in every direction.

Mark Fraser, the American I'd met in the Round-Tahiti race,

wanted to come with me. Initially I was reluctant, because his past was somewhat dodgy. He'd been a conscientious objector to the Vietnam War and had spent some time hiding out in Mexico, where he supported himself by smuggling marijuana across the border to California. His father was a well-respected banker in Aspen, Colorado, and, according to Mark, his dad had been driving cross-town in a limousine with some important clients when the news came over the radio that a twenty-eight-year-old Aspen man named Mark Fraser had been arrested in connection with the biggest shipment of marijuana ever to come out of Mexico, seized the previous day by Federal agents in San Diego. When Mark eventually got out of jail he cleansed his soul in the Pacific, taking photographs and working the ski runs in New Zealand and then making his way back up to Polynesia.

He may have been a rogue but he was a survivor too, and I found I liked him more and more. I signed him on and we set sail for the Marquesas, travelling part of the way in company with the *Mayan Five*, a beautiful Alden schooner owned by the band members of Crosby, Stills, Nash and Young.

We made good time and sailed into Taiohae Bay on the island of Nuku Hiva on a beautiful clear day. This had been Herman Melville's home when he wrote *Moby Dick,* and it wasn't hard to see where the inspiration came from. It was a dramatic, rugged island with volcanic peaks rising up from behind bays of coral, and the islanders were dark, handsome Polynesians. It was an idyllic place in many ways and, not surprisingly, it had become a stop on the hippie trail across the Pacific. Boats came and went, and there was usually an interesting assortment of people at our deep green anchorage. We used to swim between the boats, doing the rounds, until I noticed a shoal of fish going mad right behind me and realised we must have had a shark in our midst. I broke all records swimming the 150 metres back to *Maris.*

One of the more amazing boats to arrive at the anchorage was a 19-foot trailer sailor called *Super Shrimp of London.* The English skipper had painted it two shades of pink and sailed it from

Southhampton. In the south of France he'd met a beautiful girl and she'd sailed away with him across two oceans on this tiny boat. They had no money at all. In port he would work for a bag of grain which he used to make flour. And that was what he and his beautiful lady lived on — flour and fish and water.

Mark Fraser was teaching me a few tricks about living on virtually nothing too. He was a scammer from way back, but mostly he was just clever about negotiating deals. He was a good photographer, for example, and rather than pay the bill in a restaurant he'd offer to take a photo of the proprietor and his children. And in the galley he knew how to make a little fish go a long way with the use of rice and spice.

In Nuku Hiva we befriended a little boy named Albert, about nine or ten years old. His family had horses and Mark cut a deal with him for us to borrow a couple to ride over to the other side of the island. We were both taken aback when Albert brought the horses down to the beach for us. They were more like bags of bones with a horse hide thrown over the top, but they got us up over the mountains and down onto the windward coast. We walked up into waterfalls and discovered tiki statues carved into lava, and little houses with chains across the roofs for protection in the hurricane season.

When we returned the horses, we invited Albert out to the boat where we would present him with a gift for his kindness. While Mark entertained him in the cockpit I searched the cabin for something appropriate to give him. The little fellow had three fingers missing from his left hand. When we asked what had happened he said, 'Coco.' Apparently he'd missed while opening a coconut with a machete. Figuring that he'd learnt his lesson and would have some respect for a sharp object, I presented him with my German stainless steel fishing knife. I'd just sewn a leather sheath for it, and Albert may have been more impressed with that than he was with the knife. He beamed and thanked us both in his funny Franglais.

A few days later Albert's father, Etienne, paddled out to the boat in his dugout canoe and announced: 'I want you to take me diving.' We took *Maris* out to a special spot and Etienne started diving for octopus. I went down with him and marvelled at his dexterity with a spear. In no time at all we had a feast of octopus and a few fish as well.

But if Etienne and his family represented the friendly face of Nuku Hiva, the boys at the French trader's store represented the opposite. The old trader had built a kind of annexe at one end of his waterfront store and it served as the bar. He sold huge cheap jugs of wine and some pretty dodgy beer. From time to time the local boys would take the bar over and go on monumental benders which usually culminated in drink-until-you-drop sessions with 'bush beer', the home brew in which they'd put anything alcoholic they could find.

One afternoon Mark and I were ashore with a girlfriend of Mark's off one of the other boats. We'd bought our provisions and thought we had managed to avoid the crew of seriously drunk young men lounging around outside. We would leave our dinghy on the beach in front of the trader's store, and the procedure when leaving was to wait for a break between sets, then quickly push through the waves and into deeper water. The girl was in the bow, I was ready to row and Mark was in the water, about to launch us, when a couple of the drunks came down and muttered to Mark that he should get in the boat too. They would push us off.

There was a long, tense wait while more men staggered down from the bar. I suddenly realised that these guys had no intention of helping us. I watched them looking at the girl and remembered what I'd been told about romance in certain parts of Polynesia, that it's not far removed from violence. If you like a girl, you go and rape her and she will respect you as a strong man who gets what he wants. There was a distraction on one side of the boat and one of the drunks made his move down the other, whipping the girl out of the boat. Then another guy grabbed Mark. I jumped out, king-hit the man holding Mark's girl and grabbed her back. Then it was on for young and old.

As we fought our way up and down the shoreline I failed to notice that anything that wasn't nailed down was being removed from the dinghy — oars, water bottles, frisbees. Fortunately the drunks were in no condition to present a bold showing, because in strength and numbers they had it all over us. Even so, we were starting to take a few blows when an older Polynesian man came down and started shouting at them. I don't know what he said but they turned on him and that was enough to give us our break. We hand-paddled furiously away from the *melée*.

I was furious about losing the oars, so next morning I went ashore and found the leader of the drunken rabble. He had sobered up and was now full of remorse and on the slide into a ferocious hangover. I told him I wanted the oars and the other things returned by sunset or I would go to the gendarme. He grunted. Sunset came and went with no oars, so I went to the gendarme and told him my story in tragic schoolboy French. The Polynesians love a story, particularly if it involves fist fights and mayhem, so I gave him the long version. A couple of days later all our things were returned and the drunks did thirty days in the slammer. I felt a little guilty about this but Mark just shrugged his shoulders. 'Thirty days is nothin', man.' I figured he would know.

From the Marquesas we roared up to Hawaii in thirteen days, beating *Mayan Five* by five days. It was an exhilarating sail and we put into Hilo, on the big island of Hawaii, full of high spirits. A Californian friend of Mark's had flown into Hilo to meet us, and together we rented a car and drove up to the mountains, playing tourists at Kilauea Volcano, where, from a vantage point near the crater, we watched the near-continuous eruption.

From Hilo we took Mark's friend, Bruce Pine, with us across the Alenuihaha Channel and in through Lahaina Roads to the old whaling port on the island of Maui. Although, by the mid-70s, Lahaina had become something of a tourist town and had no doubt lost the frontier charm enjoyed by Mark Twain, Herman Melville and countless others, it was still very funky with a friendly waterfront scene into which we blended perfectly. We hung out, surfed, fished, sailed and worked on boats for a buck here and there. It was such an enjoyably lazy life that I tended to overlook the fact that reality was bearing down on me again. Judy was travelling to Europe with our friends Bob and Carol Lane, and they had decided to come via Honolulu to see me. I didn't know how I felt about that reunion, and my ambivalence worried me.

We finally set sail for Oahu and parked *Maris* in the crowded confines of the Ala Wai boat harbour, adjacent to the tourist strip of Waikiki. The first thing I did, as in any port, was to collect my mail. And, as in every port of call so far, the Earls had not let me down. This time, however, there was a very special letter from Kath, who perhaps sensed that my marriage was in trouble. She wrote, with the descriptive flourish of the poet:

> Autumn, Mosman Bay. Last day of April, 25 degrees, NW wind, owls asleep, rosellas at the coral flowers, Jack painting . . . Dear wandering Ian, Aloha nui loa . . . Knew you'd appreciate the crazy scene of Lahaina, more sophisticated than when we lived aboard *Maris* there twelve years ago . . . By this you have probably arrived at the best yacht club in the world [Hawaii Yacht Club] and met some of the best people . . . Give fond love to our friends [and she named about 100!] . . . Judy came to Ben's 21st birthday party looking elegant and lovely as ever. She told us she'd be seeing you in Honolulu en route to Europe. Won't that be terrific? She's really keen to be with you there, and here's hoping you can be together for a long time.

Well, we'd see about that, but I knew in the Earls I had a second family who understood the strange and compelling fancies of a man who tastes the salt air, and I knew they'd stand by me no matter what. And I really didn't know what the future held. I sat at the chart table and wrote long entries in my Blue Book, hoping that putting my thoughts on paper might clarify them. In March I had listed the priorities in my life: 'Judy, the kids, the house at Moruben Road, the future, the boat . . . What should I do? Should I go back and try to salvage the empire, or would it be better to start building again in a small way, or start again somewhere else, or even do something else? So what is the future? Is this a recession or is it a depression or just a ploy to get capitalism revitalised and healthy again? Or is it the collapse of the system? Whatever it is I have to do something and do it well. And certainly, keep sailing.'

My determination to fight to save what was left of IBK fluctuated with the latest information from the home front. In early April I was worried that the finance companies were putting the final nails in the coffin by deferring the building program. In my private chart table office half a world away, I formulated a philosophy: 'My experience tells me that when you're down the best thing to do is come up fighting and whip them off their feet so they don't even understand where you've come from . . . The problem isn't made that can't be beaten. But at the same time, how do you escape the bullshit of business? Maybe there are some opportunities in Hawaii or the US, but I have to sort out the problems in Australia first, and Australia might be in bad shape for five or ten years . . .'

Troppo? Maybe a little bit. Worried? Yes, and the more months I put between the nightmare of the collapse and my Pacific dream, the more difficult it became to confront the future.

Mark flew home to the States and I was alone again. But not for long.

> *I first met Ian on Lord Howe Island in 1974, but I ran into him again the following year after I'd sailed a 30-footer called* Illusion *up to Honolulu. I was at the marina and* Maris *came in and docked alongside us. I think we probably had a few drinks. He was a party animal, a great guy to go out on the town with. He was racing* Maris *in the yacht club races and I crewed with him a few times. We went back to the yacht club after one race and a young lady with us had a sudden urge to*

throw a pie at Bicky. There wasn't a pie available so I kindly dashed off and found a big creamy number. I gave it to her and she pied him sensationally. The Yanks cracked up. Since then, we've spent the better part of twenty years throwing pies at each other whenever we can.

Martin Farrand, yachtsman

I guess I turned into a bit of a sun-bleached layabout that spring in Waikiki. A lot of things were playing on my mind, not least of them Judy's impending arrival, and the surf and sail culture of that hedonistic resort community offered the perfect hideaway. I had *Maris* to myself, no shipmates to consider, a hundred willing playmates around the waterfront and the times were with me. People talk glowingly about the '60s, but we hadn't learnt to live the '60s dream until the '70s, and then we realised we were on borrowed time.

There was a kind of seafaring commune at Ala Wai. Most nights we'd pool our meagre food resources and barbeque on the dock, then sit around and talk story, drink a bit. We weren't very kind to our environment and Ala Wai was a mess most of the time, with beer cans and plastic wrappers adding to the waste water that ran out of the Ala Wai Canal. One day there was a handwritten notice on the marina board amongst the pleas for a ride and messages from the heart: 'Clean Up Ala Wai This Sunday. Help Us Make This Place Beautiful Again.'

We all turned up at the appointed hour with whatever buckets and tools we could find on our boats, and we did a good job too. Then the next day the rains bucketed down and the canal was full of shit again. But it felt good to have done something, and I always remembered that feeling.

In early 1975 I sailed Jack Rooklyn's boat, Ballyhoo *up to California for the TransPacific Yacht Race from Los Angeles to Honolulu. We pulled into Ala Wai in May, I think, and the first person we saw was Bicky. I'd met him around the waterfront in Sydney a couple of years before, when he'd had*

> *problems with* Maris. *He saw us tie up and he rowed straight over. Jesus Christ, that was the end of that! We played pretty hard for about ten days, then we did the sensible thing and got the hell away from him and over to LA. When we got back after the race Bicky was still bloody there! Every day he'd say he was going, but he wouldn't.*
>
> Tom Stephenson, yachtsman

Judy and the Lanes arrived in late May. I had an old Dodge that my Hawaiian mate Aka had organised for me. It was a shitheap of a thing but we had a lot of fun cruising around town in it and making daily trips to the surf at Makapuu Beach, and I had grown quite attached to it. I drove out through Ala Moana to the airport to meet them, quite excited and somewhat nervous at the prospect. Of course I had seen Judy at Christmas and things had been fine, if a little cool at times.

But since Christmas I'd changed. Put it down to the motion of the ocean, the solitude of the sea. I'd started to re-evaluate my life and there was no turning back. I'd found that I could think very clearly and constructively on the ocean, so sailing had become an intellectual pursuit as well as a physical one.

Judy and the Lanes arrived dressed for Europe. I had on a holey tee shirt and a pair of cut-offs. To be fair, they'd left the start of a Sydney winter just ten hours before, but the contrast between their attire and mine seemed to underline how far apart we'd drifted. When I saw her face I was genuinely happy to see my wife, but there was an underlying tension almost from the start. I could sense that Judy was not entirely comfortable in the beat-up Dodge as I conveyed the visitors to the plush Kahala Hilton, where suites had been booked.

I was a little uncomfortable myself when the bell boy ran out and ordered me to remove the Dodge from the hotel's driveway. I ignored him and summoned help with the guests' baggage, but when they had moved into the reception hall, I grabbed the guy and thrust the car keys into his hand. I said: 'I'm a guest in this hotel. This is my car, these are my keys, now be a good fellow and park it for me.'

The next night we had a reunion dinner in Waikiki. It started out on a high note but soon plummeted. One of my Hawaiian friends came by our table to say hello, and handed me a letter, which had apparently arrived for me at the dock in my absence. I pocketed it and thought no more about it, but this seemed to signal the end of the festivities for Judy. It's funny how, when things go seriously wrong in a relationship, it's the stupid little things you choose to fight about. Back at the hotel we fought about the letter. In fact, it was from my Austrian skiing mate, but that soon became irrelevant. Judy threw the book at me. She had a long list of grievances, both business and personal.

In the end it became a lecture — when you get home you will do this, you will do that . . . I'd had the freedom of the ocean for too long and I just couldn't take it. There were some tears and I stormed out. I remember driving down through the city to the boat with Kath Earl's words ringing in my ears, 'Here's hoping you can be together for a long time.' Doesn't look like it, Kath, dear friend.

I sat at the chart table on *Maris* and wrote a long entry in the Blue Book, concluding, 'So what am I going to do? Go home, sell up, split up, go elsewhere?' I needed some thinking space. With first light approaching I slipped out of Ala Wai and sailed around Diamond Head and into the Molokai Channel.

When I returned two days later, Judy had flown on to Europe.

CHAPTER NINE

HOMEWARD BOUND

I tossed up whether to sail *Maris* home alone. I was ready for it, and in some ways I was in the right frame of mind. Still, something made me seek company for the voyage.

When I'd first arrived in Honolulu it had still been my intention to sail on across the Pacific to San Francisco, and in that context I'd talked to Scott Coombs, a young sailor from Seattle, Washington, about crewing with me. But now time had slipped away and I knew I had no alternative but to head for home. I asked Scott if he was interested in sailing to Australia, and he jumped at the chance. Scotty was a lean, athletic young guy who'd sailed on *Maris* with Marty Farrand and me in club races, so I knew his capabilities, and I looked forward to sharing with him the adventure of crossing the Pacific.

We sailed out of Honolulu in July and made for Palmyra, one of the Line Islands about 1,600 kilometres south of Hawaii. There was a bit of a mystique about Palmyra. During World War Two it had been a major air transport base, the last line of defence in the Pacific, and there had been thousands of troops posted there. Since then, however, it had been all but forgotten, an uninhabited atoll of some fifty small islets whose central lagoon had been blasted out by the US Marines. All this had been done more than thirty years before my arrival and getting in and out of Palmyra's lagoon was a major challenge. There were coral heads, a strong current and no markers. In fact a few months before Scotty and I eased into the lagoon, the New Zealand yacht *Tequila*, skippered by my mate Darcy Whiting, had suffered a broken propeller on the coral heads and Darcy had to fashion a wooden one to get out.

Palmyra was also famous as the setting for one of the Pacific yachting fraternity's great murder mysteries. In the late 1960s it was commonplace for gypsies, pirates and hippy vagabond sailors to take advantage of the abandoned port facility, and at one time two yachts at neighbouring anchorages became quite friendly. One was a typical cruising yacht, spartan but comfortable enough for the free-spirited couple who sailed her. The other was a run-down yacht crewed by modern pirates who made their trade in dope, guns or anything going. Pirate boats at that time held a certain fascination for cruising yachties, to whom the world seemed open, free and honest. It was therefore not so unusual that the couple befriended the pirates while they shared this remote anchorage.

But then the couple was never seen again, while their yacht was later identified, with a paint job, somewhere in the Hawaiian chain. The pirates had murdered them and dumped them at the bottom of Palmyra Lagoon, then scuttled their own boat and moved on. It was the sort of pirate yarn you hear wherever yachties gather, but in this case quite true.

When we got to it, Palmyra was still an abandoned military base but now the home of a hermit called Hal, who scavenged around the shoreline for his tucker and had a ham radio setup. It was an eerie sort of a place, with jungle-invaded huts, overgrown runways and rusted skeletons of jeeps and trucks. To add insult to injury, the reef fish were contaminated with ciguetera, and we were reduced to scavenging for bush tucker along the edge of the runway with Hal. At the time the US government was considering the appalling option of using the atoll for nuclear waste disposal.

Our diet at Palmyra consisted of seabird eggs, coconuts, palm hearts and tinned food, which we rationed. As we made our way down to Samoa we started to fish in earnest, with immediate results. Our first strike was a big yellowfin tuna. We hauled the fish into the cockpit, clubbed him and cut a slab of the still-pulsating flesh. We ate this in strips of raw meat, dressed with only soy sauce and wasabi, the Japanese horseradish paste. I can tell you it was the best sashimi I have ever tasted. Second meal off the fish was tuna steaks, and then I employed the preserving technique taught to me by the Tahitians and Hawaiians. This involved crisscrossing the fillets with a sharp knife and pouring rock salt into the cuts, creating slabs of dried tuna that we could eat for weeks with rice.

We went through the equatorial doldrums, then picked up some breeze which was soon gusting over twenty-five knots and ripped our mainsail. While I was sewing it together the seas turned big and lumpy. The weather seemed set for days, so Scotty and I reefed the boat down and continued under shortened sail. On about the third day of this I woke just after dawn. Scotty was in his bunk and the steering pilot was on. The pitch of the boat and the familiar thumping seemed about right. All was well with the world. I jumped out of my bunk and put the water on the boil for coffee, then leapt back into the comfort zone to wait for it to boil.

I must have dozed off momentarily, because I suddenly became aware that the water was taking too long. I got up to have a look. The water was nowhere near boiled; the stove obviously needed fuel. Unable to buy methylated spirits in Hawaii, I had refuelled with industrial alcohol, which was quite effective but, the sellers warned, somewhat volatile. I opened up the tank, put a funnel in and started filling from a two-gallon drum.

At that moment a wave hit the side of the boat, causing me to lurch forward and spill the fuel. I hadn't noticed it, but there was still a tiny flame under the kettle from the dregs, and this now ignited the spilt fuel and the drum. Suddenly I had a fireball in my hands, spraying flames all over the cabin. I jammed my left hand over the cap to stop it, but this just blew the side out of the plastic drum. Then it was really serious. This stuff wasn't like metho, it was roaring! I had a clear moment in the midst of panic and threw the damn thing out the hatch. Unfortunately it set the mizzen sail on fire, so then we had a fire on deck too.

By this time Scotty was up and in action. We both went for the fire extinguisher, but we fumbled and couldn't get it off its rack. There wasn't time for analysing the situation; I ran on deck and grabbed the washing water bucket, observing my hand for the first time as I did it. It looked like a cooked chook! I raced below with the bucket of water and threw it at the flames. Scotty had located the spare five-gallon drum of water we kept for the life raft and was pouring it onto his flaming bunk. A few more buckets from on deck and we started to bring the fire under control. It was then that Scotty noticed my legs were on fire. He grabbed a pillow and quickly smothered the flames.

The cabin stank of burnt flesh, burnt vinyl, burnt paint and burnt alcohol. We slumped, exhausted in the middle of this mess, and I said to Scotty: 'Mate, what have we done?'

The pain took a while to register through the shock, so I pulled a couple of cold beers out and gave one to Scott. We sat on the floor of the burnt-out cabin and drank in our situation. We were a week or more off Pago Pago with favourable winds, two weeks without them, and we both had severe third-degree burns. Huge blisters were already starting to form over my legs, arms and upper torso. While I could still move without too much pain, I pulled out the medical kit to see what I could find. Heading deeper into shock by the minute, I found a tube of Dencorub and rubbed it into my wounds. It was all I could do to stop myself from screaming. For the next few hours I hung out on the afterdeck with my limbs trailing in the cool water, while Scotty ran the boat. But he was in not much better shape. I knew we were in deep shit and that it would take a lot of strength of character to sail *Maris* to her next port of call.

The next day was undoubtedly the worst. I woke to take my watch and swung my feet off the bunk. As they hit the floor the blood rushed to them, causing me such agony it took me at least ten minutes to consider my next move. As the initial pain receded, our wounds began to weep, and in the salt air and humidity there was no way we could keep them clean. We dressed them several times a day as best we could, but there was a real danger of serious infection. In my time in the Pacific I had seen people with limbs missing from staphylococcus infections which had started out as tropical ulcers. We would be heading down that road unless we had proper treatment soon.

After what seemed like an eternity we got to Pago Pago, the capital of American Samoa. The harbour there is a pretty dramatic place, filling a ravine between volcanic mountains. But sadly the American strategic occupation had caused the town to lose the magic of its Pacific neighbours. This hardly concerned us at the time, our main aim being to check into the LBJ Medical Centre without delay.

We nosed our way up to the Customs office at the end of the harbour where some officials were waiting to meet us. While Scotty edged the bow in, I stood on the pulpit with the ship's papers under my arm. I must have looked a sight, with filthy bandages over my sores, unshaven face and matted hair. With a supreme effort I jumped half a metre onto the dock, landing on a bottle. The glass crushed under the reef-toughened skin of my bare foot. I flicked it away like some unkempt Mongol warrior and introduced myself to the Customs officials.

By this time I knew enough about the Polynesian temperament to tell them the long version of the story. They loved it, laughing uproariously at the part where I set fire to the cabin, then making appropriately sympathetic sighs as I described our burns. They agreed to make arrangements for us to be treated at the LBJ Centre while we anchored the boat. Later, however, when we came ashore in the dinghy, they were nowhere to be found, and Scotty and I hitchhiked along the waterfront. A lot of vehicles passed us before a dilapidated truck ground to a halt and the driver looked us up and down before agreeing to drive us to the hospital. We climbed into the tray and bounced off up the road.

The first reaction of the doctors was to admit us until the burns had begun to heal properly. I explained that I had a boat that would get very lonely in the bay without us. In a letter that reached me in Honolulu, Jack had warned me about Pago Pago. 'Never take your eyes off the ship. Two yachts were stripped while we were there.' The doctors reluctantly agreed to treat us as outpatients. The initial treatment was quite horrifying. I watched in amazement as the doctor stripped off the filthy bandages then peeled away the burnt skin as if he was removing a pair of rubber gloves. I came out of the ward looking like a mummy.

We had a supply of ointment and clean bandages from the hospital, and on the way back to the boat we bought provisions and collected our mail. Then we anchored *Maris* way up the bay, as far away from land as we could get while still in protected waters. I was paranoid about infection and wanted to be away from the fly zone. We went right through the cabin with Dettol, then sat down and relaxed while our hides grew back.

We were anchored in Pago Pago for about two weeks and went ashore only three times, for doctor's checkups and provisions. Every day Japanese and Taiwanese fishing boats brought their tuna catches into the cannery and the story about the 'poor burnt bastards out on the yawl' must have done the rounds. We'd just about finished our stock of dried tuna when one of the Japanese skippers came aboard with a big tuna, straight out of his freezer. He also presented us with two pairs of fishermen's gloves, which were a wonderful gift because even the simplest of chores was painful to the tender new skin on our hands.

While we recovered, I had ample time to consider my financial position, since Judy had documented it, at my request, in a six-page letter. It was cool, terse and to the point:

> Hi there. One business letter as promised coming up. Construction company. Seems to be looking okay. Have asked for some figures as so far only have estimated

> expenditure for twelve months. [Finance company] says income is too hard to project. [Bullshit!] . . . Tierra [Del Fuego] dried up for the time being till they decide what to do with the remaining houses . . . Would really like to diversify and go after bigger stuff, but perhaps that should wait until you come back. Neville (IBK accountant) says we don't have enough men. My theory is if you get the work then employ the men. Lend Lease didn't go into the big league with seventeen men.

In my absence Judy had really taken the bit between her teeth in running IBK Constructions. With the rest of the empire crashing all around her, I had to admire her for that. But I was becoming deeply concerned about her expansionist plans. The group was in deep trouble and we were in no position to fund a fast-growing company. To my mind, expansion at this stage was a recipe for disaster, but she was the boss. I had forfeited my right to say no.

She went on to document the latest instalments in the carving up of Squizzy Taylor Square and Ida Pender Place under the receiver. It made terribly depressing reading because I could see that little or no progress had been made on so many fronts. If the receivers had been brought in to engineer an orderly sale of assets to recover the debt, then they appeared to have failed miserably. Very few of the properties had been sold at market value, despite the fact that the economy seemed to be improving slightly. The joint-venture profit and loss statement for the year to 30 June 1975 showed that sales of the terrace houses had raised $1,210,575, but that the 'cost of sales' had been almost as much! I knew that I would be walking back into a fairly hostile environment. Just how I would deal with that, I knew not. But I knew it would not be the same way I would have dealt with it before the voyage, for I was no longer the same person.

Judy finished on a slightly more personal note: 'Thanks for your cable. Glad your burns are healing . . . It's your mother's birthday September 1. Love, Judy.'

Finally we were in shape to sail south again, this time for the Vava'u group of islands in northern Tonga, which we made with incredible speed in following winds, passing along the coast in clear, deep water with volcanic mountains rising up beside us. When Jack heard we were at Vava'u he was thrilled. He wrote: 'How about you! "His way once made, he forward thrust outright nor stepped aside for dangers or delight."' The islands were stunningly beautiful, but Jack was right. We were on our way home and dangers or delights were out of the question.

We sailed on down to Lord Howe Island. Of course, there was a letter from Jack, his last for the voyage. Since he had been with me all the way, he knew exactly how I'd be feeling. 'I'll bet you're pretty sad, old mate, as you and your *Maris* near these shores again and the glory of this voyage draws to a close.' It was a glory no one else could share nor truly understand. But Jack did. It was the glory of accomplishment, of conquering a mighty ocean in a small boat. There was nothing particularly brave or special about my journey. I was the rookie, *Maris* had crossed the Pacific three times before! But the ellipse I had sailed had a symmetry my life ashore lacked, and I knew I had to change that. And I knew it wasn't going to be easy.

Scotty and I sailed from Lord Howe to Sydney in fifty-four hours. It was totally unplanned, but I sailed through the Heads on 16 September 1975, a year to the day since I had left them. It was a lovely spring evening and I remember looking at the city skyline and thinking that whatever I had owned in this place, I could no longer call my own. This was going to be a fresh start.

I had established new values, and written them down at great length in my Blue Book so that I couldn't resile from them. The accumulation of money was no longer my prime goal in life. I suppose that's easy to say when you've reached dizzy heights and toppled from them, but I really believed that people like Mark Fraser had taught me a new approach to life. I knew now what was important to me, and it wasn't the size of the house, the car or the bank balance. At the same time, however, I had to face the realities of my financial situation. I owed a lot of money and I knew I could never pay it back unless I was able to maintain a certain mobility in business. Basically, I needed the sharks to back off and give me some room to move, to start negotiating my way out of the mess while earning enough money to live. I was determined that they would, because I'd been brought up to believe that going broke was unacceptable. Only losers cried poor bugger me and walked away from their financial responsibilities, and I wasn't a loser. Fifteen years later, it was interesting to watch, with a degree of detachment, the same cycle of events, how the sharks moved in and how the high-rollers of the '80s reacted to their falls from grace. I was never in their league, of course, but I'm bloody glad I didn't act like a Bond or a Skase. There might have been a few jokes about *Maris*'s gold keel, but I was back and I knew what I had to do.

The first step was to start a new company, IBK Builders, which I ran from home at Moruben Road while Judy ran the affairs of IBK Constructions and the rest of the group from our city office. In business I had decided to become a minimalist. All I needed was a phone, a truck and a bag of tools and I was away. I called in a few favours to get things started, but the reputation of the building teams I had put together in the past was sufficient to get IBK Builders plenty of contract work.

By December, IBK Builders was in good shape, so I took up an offer to crew on *Apollo II* in the Sydney to Hobart race. Then I went across to New Zealand to do the Dunhill Cup on *Ballyhoo.* I stayed with Marty Farrand in Auckland and we relived the Hawaiian summer, throwing pies at each other and having a jolly old time.

I don't know if this was the last straw with Judy or whether our relationship had long passed that stage, but when I got back to Sydney the situation was untenable. It was hurting us both so I moved out. As far as I was concerned, our ten-year marriage was over.

CHAPTER TEN

STARTING OVER

By early 1976 I had enough capital to get back into the real estate market. I lined up some private finance and bought two adjoining houses in Paraween Street, Cremorne, just a couple of inlets around from Mosman Bay. I moved into one and the Bilge Rat moved into the other, and we started to do them up. The Rat, nothing if not an improviser, had been living in a cave adjacent to the marina. He was quite comfortable there and actually disputed the wisdom of moving, rent-free, into the new house. I said: 'It'll be hell, Bilge, but a job is a job.'

Although it was a cautious re-entry into the minefield of Sydney real estate, I had a good feeling about the Cremorne houses, as well as a firm grip on the handbrake. There was no way I was going to overgear again, but a small stable of rent earners would put me in better shape to deal with the problems. And the problems were continuing to mount. Suddenly the reissue of my builder's licence was in jeopardy.

All these years later, when so much has happened in my life and so many honours have been bestowed upon me, I still tingle with rage at the thought of what the sharks from the finance companies, through the good offices of the Builders' Licensing Board, tried to do to me in 1976 and 1977. To me it was incomprehensible that my means of livelihood should be taken away because some of my companies had failed. To this day I regard the unfortunate collapse of 1974 as something that was largely beyond my control. I did nothing illegal, nothing morally wrong, nothing to bring the building trades into disrepute, and certainly nothing to hurt the tradesmen who worked for me. In fact, they were the first to be paid from whatever cash I could muster.

I had played the gearing game and lost, like so many other developers who created work in the building industry in the 1970s. Maybe I was naive, maybe I got too greedy, but for these heinous crimes it was now suggested I should surrender my right to earn a living because I was 'not a fit and proper person'.

If any good came out of this attempt to stop me from working, it was that I became aware of the huge battles looming in the courts and resolved to fight with everything I had. For starters I sacked my lawyers, who seemed to be getting nowhere, and eventually got onto a firm called Robert Burge & Co. I met with a lawyer named Phil Esplin who listened intently as I outlined my position. Esplin took copious notes as I spoke, interrupted occasionally to clarify certain points, and when I had finished, sat back in his chair with his hands behind his head and declared: 'There's a hell of a lot of work to do here, but I think we can win a few battles.'

I liked Esplin's style. He was aggressive and quick and you knew you were dealing with a professional. But, as it turned out, he was not to be the man in my corner of the ring. That honour went to a young lawyer named Bruce Hambrett, who had Esplin's fighting style plus the raw energy of youth. Bruce immediately went to work on the biggest problem, the reissue of my building licence, without which I couldn't pay my legal bills, let alone begin to retire a mountain of debt.

The Builders' Licensing Board was one of about five battles he had going, but it was the big one. They were all quite daunting cases because the opponents were big and they had deep pockets. Ian wasn't very well organised and the paperwork was lying around the place in boxes everywhere. But he was pragmatic and very supportive. I recovered some money for him in another case and he handed me back the cheque and said, 'That's the fighting fund.' I knew then I had a client who was serious about winning. He wanted me to work miracles, but he was prepared to pay for them.

I never thought what Ian had done, in going away sailing, was in any way irresponsible. You can't go sailing forever. He was always going to come back. He'd taken time out to analyse his position, and he'd come back determined to beat his problems. We fought and we eventually won them all, but the sweetest victory of all was the Builders' Licensing Board. That was the bureaucracy taking a swipe, and we swiped back.

Bruce Hambrett, lawyer

The case against the Builders' Licensing Board slowly began to turn our way, and the day I got my licence back, reinstated on the spot, the District Court judge, in his summation, delivered a stinging attack on the Board. It wasn't much compensation for so many months in which I had been unable to work, but to me it was a matter of pride. I was proud of being a builder, and I was a bloody good one. No one could take that away from me. I'm not a particularly emotional person, but I left the court that day with tears in my eyes.

Bruce Hambrett went on to negotiate deals with each of my creditors, one by one. As time went by, things became a little easier, for both Gordon Stewart and me, for we were the ones left with the personal guarantees. We started to rack up victories in the courts and at the negotiating table, and I think one of the reasons for that was that common sense began to prevail. As the major players for the finance companies moved on to be replaced by new faces to whom this was just another 'unresolved matter' on their ledgers, the sting went out of the attack. These people could see that there was an offer on the table. They were just doing a job, I was fighting for my life. In most cases they took the offer, and by

1984, after a decade of fighting, I was able to bury the memories of the crash of '74.

During 1976 Syd Fischer launched *Ragamuffin III* and I again crewed on her during the season. At the end of the summer we won selection in the Australian team for the 1977 Admiral's Cup, a team which also included *Superstar*, from Victoria, and Jim Hardy's boat *Nyamba*. The Admiral's Cup is recognised as the world's premier ocean-racing regatta, held every two years in the unpredictable seas off England's southern coast. It is a team event in which each country's points are tallied over two long and two short races, much like our own Southern Cross Cup, and this creates a terrific camaraderie between crews. This was particularly so in '77, when we had a marvellous bunch of sailors who really put their backs into the task — although we didn't win — and their spirits into the accompanying social agenda.

> *I guess I got to know Ian better that year than I ever had before. I was on* Superstar *and he was on* Ragamuffin, *but we managed to spend a lot of the time ashore chasing women together. He was very good at it. We were staying at a guesthouse in Cowes on the Isle of Wight and there was supposedly a team curfew, but we managed to break all the rules and get away with it. One night at the Regency Club, Ian managed to chat up a pretty young waitress and headed off into the night with her. I found a young lady too, and late in the evening we made our way along to* Superstar, *which was tied up next to* Ragamuffin. *After a couple of hours of dalliance, we crept off, treading lightly over* Ragamuffin*'s deck.*
>
> *Suddenly a voice roared: 'Who the fuck is that?' The lights went on to reveal a less-than-amused Syd Fischer, our team captain, clambering out of a quarter bunk to investigate, only to discover in the opposite quarter-berth a rather sheepish Ian and his young lady.*
>
> Tom Stephenson

Ahoy Ian — Everyone rooting for Ragamuffin. *I wrote another letter to you illustrated with a variety of couples, all rooting in various ways, for you and the crew, but considered it would not be in good taste at Cowes.*

Jack Earl

When I returned from the Admiral's Cup, still disqualified from contract building because of the licence case, I turned to real estate speculation again, this time with a derelict old stone house in the Middle Harbour suburb of Cammeray. No 2, Vale Street had a roof but no floor, no windows, no power, no water, no sewerage. It represented a huge challenge to a renovator and a fantastic opportunity for a speculator. I moved in, virtually camping in the shell of the building, and began to work on it.

Some months before, at a party at the home of my Crash and Burner mate, Jim Scaysbrook, I had met a wild and wonderful girl named Carole Laird. She worked as a camera operator at the Hanna Barbera studio in Sydney and, like me, rode a Honda 4. She was blonde, green-eyed, vivacious and had a great sense of humour. We fell in love almost immediately. I was still married with a divorce in the offing, and the last thing I needed to do was to further complicate my life, but we threw caution to the winds and Carole moved into the Vale Street camp with me.

Sadly, in November 1977 Judy put IBK Constructions into liquidation. I felt bitter disappointment about this. Her expansion plans had not worked and I felt she was throwing good money after bad, but she stuck to her plan and this was the end result. For her part, she probably felt that I hadn't been around when I was really needed, that I had let other things — like my determination to be my own person — stand in the way of a successful business. And she was right.

On New Year's Eve 1977, I sat at my chart table on board *Maris* and, as had become my custom, I listed in the Blue Book my priorities for the year ahead:

> Aims for 78. Go to New Zealand. Sell Paraween Street houses and cash up. Plan and renovate Vale Street or sell. Get my gear out of storage. Win the solo race. Look at my structure with the lawyers. Build pigs for profitability. Get off the finance company hook. Be a better Australian. Get divorced. Do the Lord Howe Island race.

Four days earlier I had sailed out of the Heads again, bound for Lord Howe Island, then on to New Zealand for the Trans-Tasman singlehanded race from New Plymouth to Mooloolaba in Queensland, but before I get onto that, I should explain a couple of my New Year resolutions. Since I'd moved out of Moruben Road in early 1976 I'd had my furniture in storage and I was beginning to forget what it looked like. (Not too flash, as it turned out.)

Just before I'd left on my Pacific voyage in 1974, Jack designed and the Bilge Rat and I built a short, stubby dinghy for *Maris*. Scotty Coombs later christened the prototype 'Little Pig'. These dinghies were cheap, easy to build and fun to use, so we'd developed a small market for them. 'Small' was the operative word because, as my mates Cec Quilkey and Tommy Stephenson will attest, boat-building can be a nice, slow way to go broke. But I didn't know that in 1977, and I'd lined up a few club contracts for them.

The Trans-Tasman was to be my first single-handed race, so to prepare for it I sailed solo to Lord Howe. Jack Earl's advice was to initially stick to the coast while I settled the boat down. I did this and spent a night in Port Stephens and then struck out to the east, only to run into the mother of all gales the next day. I remember sitting in the cockpit watching a massive green roll cloud building in the south. There was a wind off the land and a swarm of insects suddenly appeared. As I sat there beating them off, the gale hit, breaking the self-steering and pounding me for twelve hours. So this was the world of single-handed sailing! Jack had, as usual, been following my progress and a letter awaited me at Lord Howe.

> Tonight's ABC TV showed a magnificent film taken from the air of some of the leading Hobart race yachts, bashing close-reefed into heavy seas. They looked terrific, working to weather under such wild conditions . . . Are you away out there running off this gale?

Before he mailed his letter, Jack got my radio position and added a much-relieved PS: '*Maris* 32.26 South, 154.10 East, going well. Got that at 1100 by phone, asked them to send the typed message . . . All relieved to know you are okay.'

I spent several weeks on Lord Howe and Sally and Pip flew over to join me. We had a wonderful holiday, fishing and swimming most days, with *Maris* anchored in that beautiful lagoon. Sally was nine now, Pip not far behind. I regretted the fact that I was so often not around when they were little, and tried to make it up to them.

I crossed the Tasman again in early February, making the now familiar landfall at Cape Maria Van Diemen. It was an uneventful voyage, except that *Maris* had developed a leak. It was something I could live with while cruising, but I was dead serious about winning the Trans-Tasman and I wanted to ensure that *Maris* was in top running order. I found a boat yard in Opua, just along from Russell in the Bay of Islands, and went to work with a Maori shipwright. We worked day and night and fixed the problem just hours before Carole flew in to Auckland to join me.

Together we sailed back up the coast, stopping at anchorages like Tombowline Bay, right at the tip of the North Island, and a place local fishermen had told me about called Rock 'n' Roll Bay. The weather was good, the seafood plentiful and the coast stunning. We took our time, then had to battle headwinds down the west coast to New Plymouth. The rules of the race stated that you had to be in New Plymouth two weeks ahead of the start, and I was a week late. It was a pretty stupid rule anyway, and I wasn't overly concerned. But the race bureaucrats apparently were. They wanted to slap a huge penalty on me, but I appealed against the decision and won. I'd had my Honda 4 shipped over to Auckland and Marty Farrand rode it down to New Plymouth for me. The boat was well prepared, so Carole and I jumped on the bike and went touring while the bureaucracy calmed down.

> *It was ridiculous. He was only late because he was jerking around. He was in all sorts of shit for no reason. Ian always has to push it.*
>
> Martin Farrand

There were quite a few Australians in the race and one of them was my Mosman mate and rival Peter 'Mouse' Mounsey. Mounsey had been taunting me for months, calling *Maris* a 'semi-submerged log', an unflattering reference to her Tassie oak keel and her heavy displacement. Mounsey himself was racing *Breakaway*, a Swanson 36-footer, and while I'd been on Lord Howe Island he'd been preparing for the race in Mosman Bay. My spy on the hill, who never missed a thing, duly reported: 'He's taking this very seriously, running to get fit . . . working on double pole spinnaker gear for one man handling. Said he would keep to a rhumb line course . . .'

I had a tee shirt I'd bought in Hawaii that had a cat playing a guitar on the front and singing, 'I love dem little mousies, mousie's what I love to eat, bite their little heads off, nibble their tiny feet'. I took to wearing it around the dock. Mounsey retaliated with a tee shirt that declared: 'I'm a little bickie muncher.' It was all pretty silly stuff, but the case of scotch we wagered on the outcome was deadly serious.

The race began early in April in light conditions that have never suited *Maris*. Soon Mounsey, whose boat revelled in light breeze, was on the radio telling me he was eighty kilometres in front. Another of the competitors, Terry Rogers, later told a story about this stage of the race which he insists is true. Terry was enjoying his watch on deck in the fair weather, polishing the golf clubs he always kept on board. He heard a voice hail him and stood up to see a large fishing boat approaching him from astern. 'Where are you from? Are you alone? Do you need any help?'

The skipper seemed most concerned, but Terry allayed his fears. As the fishing boat pulled away, Terry walked along the deck with his three-iron in his hand and called: 'There's just one thing. If you see my ball, give me a yell!'

As we got up the rhumb line the weather started to change. Tropical Cyclone Hal had been hovering in the Coral Sea and seemed to be dissipating, but it had changed its mind and was now strengthening and pushing south to meet us. The breeze built and the glass dropped, and soon we were in hurricane conditions.

Jack Earl sent me a message when he heard the news:

> *Ahoy Ian — Through the days and nights I keep a mental picture of you and* Maris *trimmed to around 300 degrees, plunging and tossing while you reset the sails and adjust the wind vane, take a swig of rum, a bite of corn horse sandwich, and catch forty winks between alarm clock signals, radio weather watches and an eye out aft for competitors. Mounsey will be hard to toss. Here's hoping you can lay Lord Howe and sight it on your port hand with a fair wind so you can relax passing Elizabeth Reef . . .*

In fact I sailed past Lord Howe Island without even sighting it. I spoke to the weather station on the island and they were recording seventy-three knots. The seas were huge and the ocean surface a smoking mass. In those conditions I was worried about Elizabeth Reef which, with its sister reef Middleton, had claimed so many boats over the years. My course took me between Lord Howe and the southern end of Elizabeth, about 240 kilometres to the north. I reckoned I had it right by my earlier sextant shots, but in a breaking sea you'd never even know you were on the reef until you heard the sickening grate of hull on coral.

On the positive side of the ledger, *Maris* was surfing beautifully down the faces and I felt I had the boat under control as she roared along. And surely I had given Mounsey the slip by this! The weather let up a bit and I was delighted to hear that Mounsey was jammed against the coast near Moreton Island, having to tack to get to the finishing line. *Maris*, coming through wide with a following wind, surfed all the way to Caloundra.

As I crossed the finish line I hailed: 'How many boats?'

'Only one.' That was New Zealander Pat Costello on *Chubasco*, a modern and pacy nine-metre sloop. Pat had taken line honours and won his division, while I was second over the line and winner of the ten-metre division. Pat had completed the course in eight days and thirteen hours, knocking almost a day off the previous record. I finished in eight days twenty-three hours, also well inside the record.

Some of the gloss was taken off a great race when I was served a summons the moment I stepped ashore at Mooloolaba, but by now I was used to the fact that anything goes in a continuing legal battle, and tried not to let it spoil the celebrations with Carole.

The next day, as Pat Costello and I went out in a motor launch to meet third-finishing Mounsey, reports began to filter in about the other competitors. Two boats had sunk. On one of them the skipper had unscrewed his wooden leg and waved it from the life raft at a rescue boat. New Zealander Bill Belcher, 65, a previous winner, was missing at sea. His wife, Aileen was distraught. In fact, however, Belcher had put his boat up on Middleton Reef at the height of the blow and had abandoned her. He was to spend eighteen days in his life raft before being miraculously rescued east of Fraser Island.

Back in Sydney I got stuck into restoring the old stone house in Vale Street. A New Zealander called Jerry Finlay happened into town and offered to help me. He was the brother of Will Finlay, who'd crewed across the Tasman with me in 1972. Jerry was a heli-mustering pilot and he'd run into power lines and nearly killed himself — unconscious for eight days. Now he was on the mend, so I gave him a room in the house and he worked a few hours a day.

Carole had become very domesticated, despite the lack of amenities at Vale Street. She saw it as some sort of country retreat within the city, and filled our backyard with chooks. She also bought a bull terrier/cattle dog bitch, Tasman, which systematically worked its way through the chooks. We had an outdoor bath which was lovely on summer evenings. One time Carole filled it to the brim, ready to luxuriate in it. She went back inside and when she returned she found a boiled fowl floating in it. Tasman stood by the bath licking her chops, waiting for the delicacy. Eventually the dog got all thirteen chooks. The only reason we kept her was that Alf was in love with her. Alf by now was old and crotchety, and I figured I couldn't break his heart at this point.

Before 1978 came to a close I had fulfilled most of my New Year's resolutions — I still wasn't divorced but I had sold the Paraween Street houses and made a good profit. Early in 1979, with its restoration complete, I sold Vale Street for $180,000, making $100,000 profit. This, combined with the return of my builder's licence, plenty of work and the gradual easing of my debt battle, meant that I could look forward to easier times ahead. I bought five kilos of gold and resolved to compete in the single-handed Trans-Pacific race from San Francisco to Kaui.

Unfortunately this upturn in my business fortunes did not coincide with a happy time in my private life. Carole and I had been together for a couple of years now, and she was clearly looking to settle down. I loved her, but this was not what I wanted in my life. I had to be free, had to have the love of a woman who could understand and accommodate my need, my selfish desire to sail across oceans. Carole tried, but in the end she just couldn't. We began to drift apart, and eventually separated.

In the early 1970s at Squizzy Taylor Square I'd met a friendly young girl named Judy Humbley who worked for John Singleton at SPASM. She took an immediate shine to Alf, and later, I suppose, to me. We became good mates and after Judy and I split up, she started doing the IBK Builders' books for me. We'd see each other for a lunch or drink from time to time, and I always thought of her as someone I could depend on. I suppose it was natural that I would gravitate towards her again after Carole left, but this time there was more. I think there is probably nothing more solid than a love born of friendship, and this one was about eight years in the making.

Unlike Carole, Judy well knew the pitfalls of loving a man who loves his boat above all else. When I told her I was going to do the Trans-Pac and would be gone for most of 1980, she took the news stoically and offered to mind Alf. I wrote in the Blue Book:

> Little Judy steady as a rock and she loves you to bits. And what a pal she is! Just a few jobs now and you can get on the first available southerly front and get to your beloved Lord Howe Island. See if you can tempt the Mouse to race you to San Francisco . . . Anyway, take it easy, be 100 per cent professional, take no unnecessary risks and look after your best girl *Maris*, your longest love to date.

CHAPTER ELEVEN

FREE AGAIN

Off I went, free as a bird, down the same old Pacific sea road. Lord Howe for Christmas with all the old mates, then on across the Tasman with Terry 'Holly' Schick from Lord Howe crewing.

We encountered a fierce gale in the Tasman and I wondered how the other boats would be faring, making their way home from the Sydney to Hobart race. The bad weather stayed with us all the way to New Zealand, and as we approached our landfall, I realised that I didn't have my charts for this coast. I dug out a road map and plotted the Kaitaia aero beacon on it, and we were able to close the coast with some confidence. But as it got close to dark I told Holly we would have to put to sea and try again in the morning. Just then the clouds broke a little and I recognised the distinctive coastline of Cape Karikari. That meant there were rocks between us and Port Manganui, but I felt confident we could clear them in the last of the light and make the inlet. We did, but not without some anxious moments when the weather closed in again.

I'd put in at Manganui several times before and knew the publican. At times, the poor man's wife would be indisposed and couldn't make it to the dinner table, which was a bonus for visiting yachties, who got to eat her meal. When I came down from my shower, sure enough the publican pulled a plate of chops and vegetables out of the oven and said, 'Here, the missus has gone again, you might as well eat this.'

As I ate the food and washed it down with a cold beer, the publican sat with me and shook his head. 'Glad to see you safe in harbour this weather. Been enough tragedy.' I didn't know what he was talking about. He explained that an Air New Zealand airliner had ploughed into Mount Erebus in the Antarctic, killing all on board, including some yachting mates well known in these parts. I had barely taken in this terrible news when he added: 'And now there's boats missing in the gale. *Smackwater Jack* bound for Auckland and no radio contact.'

I felt sick and dizzy. I knew Paul and Alison Whiting from *Smackwater Jack* and I knew that another mate was to crew with them in the Sydney to Hobart. Was it Marty Farrand? I made some phone calls and discovered that Marty was safe at home in Auckland. It was my other young Pacific shipmate, Scotty Coombs, who was somewhere out in the Tasman with the Whitings.

That night I drank some rum and shed a few tears as I wrote in the Blue Book: '21 January 1980, Port Manganui. Pam, Harry [my mates who died on the *Erebus*] and the others all gone. Scotty, you better be okay. Keep going, old mate.'

But he was gone already. A bunch of us chipped in the next day to charter a search plane, but it was hopeless. *Smackwater Jack* had been claimed by a gale so intense that even the mighty *Condor of Bermuda* had hove to for a couple of days. Nothing was ever found of the *Jack*, nor were the bodies of her crew ever found.

From Manganui we sailed south and there was mail waiting for me at the Russell post office. A letter from Jack, postmarked 9 January, close to the date *Smackwater Jack* disappeared, and certainly some time before the alarm was sounded, said in part: 'I had a terrible vision of Scotty just today, of him in deep trouble, drowning and being overwhelmed in a smother of foam. I was so shaken that I rowed ashore and went up to the compound to tell the family about it.' For all the years I knew Jack Earl he had seemed to me to possess a second sight. He was always right about the conditions that I sailed in, and, as it turned out, he was right about Scotty.

Holly and I sailed on up to Tahiti in mostly fair weather with easterly breezes pushing us along nicely. It was a good sail, if not a dramatic one, which was just what I needed. Many times on that voyage as I kept my watch in the cockpit, I pictured Scotty clambering around the deck with his burns covered in dirty bandages, never complaining as he rubbed his wounds painfully against ropes and wires. He was a good sailor, Scotty, and a fine young man. To this day I feel that his loss was such a waste of a life full of promise, but this is the game we play, I suppose.

Judy forsook Alf and flew up to Papeete to meet me, which was wonderful at that stage of our relationship. I was able to introduce her to Rodo Williams and my other Tahitian friends and give her a feel for the kind of life I was addicted to. If she couldn't share all of it, she could at least understand it.

Holly flew back to Sydney and Judy followed soon after. I was alone again in Papeete and seriously considering sailing on to Hawaii single-handedly when I was approached by a girl named Lynn Joesting from Chesapeake Bay, Maryland. She had signed on as cook in Honolulu on a boat that was now tied up near me. As sometimes happens in the Pacific, when the salt air and the scent of frangipani work their lusty magic, the skipper had read a great deal more into the crewing arrangement than Lynn had. She was there for the sailing. When she repelled his advances, he turned nasty, making her sleep in the cockpit and generally treating her like unwanted baggage. She needed to get to Hawaii.

I wasn't too keen, but I sympathised with her plight and eventually told her to go and make the skipper an offer. I would take her if he provisioned *Maris* to my specifications and paid me $500. (This was a lot cheaper than forfeiting his crewman bond.) Lynn came back an hour or so later with her bag and a smile from ear to ear. 'He went for it!'

We stocked *Maris* with wine, cheese, French bread and tropical fruit and pulled out that night. The skipper was on deck as we slipped our lines. I couldn't resist. I called out, 'Thanks for the stores, old mate. And the money . . . and the girl.'

Lynn was a good sailor and a good shipmate, and *Maris* made good time against the trade winds and we put into Honolulu in mid-May and I had the boat hauled out and anti-fouled in preparation for the long voyage to San Francisco. Jack wrote, greatly concerned that I was not leaving myself enough time to prepare for the single-handed race. 'You are not coming up for much air. The start for your solo back to Kauai on the 15th is very close. Try and rest, old mate, and EAT!'

As it turned out, I had a fast run across to the West Coast and arrived in San Francisco Bay with a week to spare. I stayed with friends of Jack's, Ron and Nancy Wise, in Sausalito, while *Maris* lay moored in Pelican Harbour marina, about the plushest in town. Ron Wise was a New Zealand expatriate, a shipwright and sailor. Any mate of Jack's was all right by the Wises, and on subsequent visits they would just hand me the key to the spare room. During the week of preparation for the race, I did as Jack suggested — slept and ate well — but I also made time to explore San Francisco Bay on *Maris*, sailing under the Golden Gate Bridge and out to the Potato Patch, where the sea turns blood red. I was intrigued by the fog, the colours of the sea, the birdlife . . . it was all so different.

The race got underway on 15 June with considerable interest within the fleet as to how a heavy-displacement boat like *Maris* would perform, in mostly light air, against the new ultra-light-displacement-boats (ULDBs). Since *Maris* had been built in the late 1950s, with a dead weight of seven tonnes, light displacement boats had become the norm for Pacific ocean racing, then came the ULDBs, which were still quite a novelty in 1980. The difference between them and *Maris*, I'm afraid, was about 500 cement bags.

The fleet, of mostly little-known boats, varied tremendously, with only five out of thirty-eight custom-designed, and not one purpose-built for the Trans-Pac. This resulted in one boat, the 55-foot trimaran *Crusader*, sailing away from the fleet on the first day, not to be seen or even thought about again until Hanalei. Skipper Michael Kane had targeted the Ostar multi-hull race, but had run out of money for the campaign and had entered the Trans-Pac, for all intents and purposes a mono-hull race. He set a course record of just under eleven days, but it had little to do with our race.

We had twenty-five to thirty-knot breezes for the first couple of days and I set *Maris* a course to skirt the roving North Pacific high, whose centre is a zone of little or no wind. Unfortunately I cut the corner a bit and landed in a calm, which put paid to my chances of shaking up the ULDBs. They ran like a shower of shit around the high, while I laboured away in doldrum conditions. In the mid-Pacific, conditions improved somewhat, but I still could only manage eleventh out of twenty-seven finishers in eighteen days and eighteen hours.

The winner on corrected time was *Sanderling*, a 24-foot Golden Gate built in the early 1950s from a 1930s San Francisco Bay design. *Sanderling* was probably the least expensive boat in the fleet, carrying few sails and very little equipment. But her skipper, Bob Counts, from Sausalito, was a no-nonsense sort of a bloke who knew the Pacific high and knew his boat. Before the race he had said: '*Sanderling* is right for this race because her full keel and small rig allow one man to sail her close to her full potential.' He was right, and it was one in the eye for the ULDBs.

After the parties in Hanalei, I sailed back down to Honolulu, put into Ala Wai and got onto the club race circuit for a few weeks. During this time I received some sobering news from home. My Crash and Burner mate Zoran Janjic sent me a telegram: 'Gold fell $200 stop Alf died stop Carole married stop.' There was a lot of the black humour of the Crash and Burners in this telegram, but that didn't disguise the fact that my great mate of fifteen years, Alf, was gone. In fact Judy had had to put him down, an act that nearly broke her heart.

I quickly made arrangements to sell my gold. As for Carole, I was happy for her, but I'd squared up to the fact that we had different goals in life, and that chapter was long closed.

It was great to be back with my Hawaiian mates again and, just as in the summer of '75, I found it hard to drag myself away.

Eventually it was Holly Schick, who had shown up out of nowhere and encouraged me to slip the lines and head home. We followed our line from '75, down through Samoa and Vava'u, then over to Lord Howe Island in time for Christmas with Judy, then home.

Back in Sydney I moved into Judy's Kirribilli terrace, which is where we've been ever since. In '75 my homecoming had been full of trepidation and uncertainty about the future. Five years later, as much as I was addicted to the cruising life, I felt glad to be home with family and friends. I'd just turned forty, my girls were about to become teenagers. Maybe I was growing up at last.

But there was one new sadness in my life. In my absence my father had grown very frail. Now in his mid-seventies, he had outlasted the lifespan predicted for the Burma POWs by army doctors by a good twenty years, and he had outlived many of his wartime mates. A couple of years earlier, Mum and my sister Ann had talked him into selling Gillingwood and moving into a terrace house in Edgecliff Road, closer to the city. He was not keen because, in his mind, terrace houses would always be associated with his hand-to-mouth existence in Melbourne during World War One, but when my mother found a lovely Victorian house she liked, and I had pronounced it of solid construction, he said softly: 'That'll be fine, dear. You're steering the boat now.' And he was right, poor old fellow. He passed away in 1981, still enjoying a smoke and a drink to the end.

We'd had our differences, George and I, and he hadn't always embraced my business decisions or my pursuit of pleasure in the Pacific. I'm sure nothing would have given him greater pleasure than to hand the firm of G.A.C. Kiernan over to his only son and to have watched it flourish in his dotage. But I think he understood why I did what I did, and I know there was always a mutual respect.

The building business was as flat as a tack, but the building company, now known as Alfred Builders in memory of my old blue dog, scored a couple of juicy contracts. One of them was to convert the old Bonython Galleries in Paddington into headquarters for Mojo, then Australia's fastest-growing advertising agency. The principals, Alan Morris and Alan Johnson, were easy-going blokes who liked a lunch and a laugh, and they gave me an open brief. I was also invited by the National Trust to tender for the restoration contract on a streetscape of nine miners' cottages at Lithgow, the first town over the Great Dividing Range. I loved the promise of what I saw, submitted a proposal and sailed back over to Lord Howe Island for another glorious Christmas. We were just getting into the swing of things on the island when I received word that Alfred Builders had won the contract.

Restoration work was not particularly lucrative, but I likened it to single-handed sailing, in that once your reputation was established you were always recognised as a specialist, and always given first call. It was also the kind of building challenge I couldn't resist, involving complete immersion in the history, culture and craft of a group of people. But I knew I would have to be careful. When you fall in love with a job, it's easy to let costs run away from you and do your arse.

I put together a crack team of tradesmen — carpenters, plumbers, bricklayers, electricians, the lot — and we set up camp in Lithgow Monday to Friday. The cottages were in shocking disrepair, having been rented cheaply for years, and it was therefore a fairly rough camp. But the blokes didn't seem to mind. There was great camaraderie. We'd drive up there on Sunday nights and home again Friday nights, and in between we worked long and hard. The Federation detail was intricate and extremely difficult to recreate, but all the guys seemed to thrive on that challenge. Of course, we made time for the occasional drink and chin-wag with the locals in the pub.

I had a wonderful bricklayer named Barney, an older bloke who always wore a King Gee navy shirt and matching trousers. He had a shock of white hair and he smoked constantly. One evening I was sitting at my desk when one of the other bricklayers raced in. 'Quick, I'm worried about Barney!' I followed him into the brickies' cottage a couple of doors down and poor Barney was stretched out on his camp mattress, looking as dead as a maggot. The previous night the men had put in a huge one at the pub, and I'd had to wake them up at six for our normal 7 am start. When I walked into the brickies' place, Barney was sitting up on his mattress looking very grey and hungover, rolling his first smoke of the day. 'Christ, Barney,' I said, 'You look crook.' Twelve hours later he was dead.

I gave up smoking that day, cold turkey, twenty a day to nothing. For two years I even dreamed about unopened cigarette packets.

Because it was politically controversial — Lithgow Council had originally wanted the cottages demolished — the job developed a high profile which led to me being approached by a woman named Iris Paradines, who worked for the National Parks and Wildlife Service. She was in charge of organising the rehabilitation of Hartley Village, an historic settlement back towards the foot of the Blue Mountains, using mostly RED (Regional Economic Development) scheme workers. This was another wonderful project, and I took to spending a day or so a week there, doing the simpler jobs by myself. I was enchanted by Hartley, by the sense of the past I felt there and by the wonder of the coarse but practical construction methods of the pioneers.

We got into some quite complex underpinning and stabilisation of these fragile buildings, and I knew I needed the practical skills of a good full-time foreman. I drove down to Mosman Bay and bought the Bilge Rat a few beers at the Rowers and asked him if he was interested. 'Why not?' he asked back.

I bought him some new overalls and a few simple cooking implements and set him up in one of the Hartley cottages. On Fridays I'd come back to Hartley to inspect progress and take him home for the weekend, but the Rat rarely wanted to come. 'I like it here,' he'd say defiantly, house-proud in his shell of a cottage.

In Nuku Hiva in the Marquesas in early 1975 I had met a big, loudmouth yank from Florida named David White. He was doing a world cruise on a Westsail 32 and, over a few beers, was liable to wax eloquent about his vision for a single-handed race around the great capes of the Southern Ocean, the ultimate test for a solo sailor. He called his race 'The Ultimate Challenge — Around Alone'. I could share this dream, even though at that stage I had yet to sail solo across an ocean, and I filed it in a back compartment of my brain, where it was to remain for half-a-dozen years.

It re-emerged in 1981, when I first heard about The BOC Challenge. BOC, the huge British-based multi-national gas company, had taken David White's concept and were set to sponsor the first race, starting from Newport, Rhode Island in August 1982. The race was to be run in four legs, with recovery periods between each, over 43,000 kilometres of the most testing ocean water in the world. I wanted to be a part of it, desperately! But I was realistic enough to realise that I didn't have a hope in hell of making the eight-month voyage without major sponsorship and a purpose-built boat, and I already knew how difficult sponsorship was to come by. The idea of sponsorship had first occurred to me when I decided to contest the Trans-Tasman Single-handed in 1978, then in the lead-up to the 1980 Trans-Pac I had made a concerted effort to attract sponsorship for a 56-footer I wanted to build. Nothing came of these plans, but now it was a new race, a new ball game.

I announced myself as an Australian starter in this historic race and began to bang the drum around town, looking for sponsors. A newly-reformed smoker, I even turned my back on this pledge — in a sense — and put what I thought was a totally convincing proposal to R.J. Reynolds, the company which made Camels. By early 1982 I had enough rejection letters to paper my walls and I realised I would not be skippering a purpose-built racer. Well, bugger it! Gallant little *Maris* would conquer the capes! In Jack Earl, I had the patronage of one of Australia's best-loved sailors; between Singo and Mo and Jo, I had connections with the top advertising agents in the country; and I had proven credentials as a winning solo sailor. But still I could not raise the $60,000 I needed to do the race.

Having shouted my intentions from the rooftop and having

paid the entry fee, I sat quietly and solemnly at the Sydney press conference called to announce details of the race. I offered my congratulations and support to the only Australian entrant, Nev Gosson (who had won $50,000 sponsorship from the new Pier One development), but I was bitterly disappointed.

When Gosson sailed out of Sydney for the race start in Newport, I took *Maris* out on the Harbour to see him off. As I left race headquarters at Pier One a young blonde woman came racing down the wharf, yelling at me to wait for her. She jumped aboard and introduced herself as Kim McKay. She said she was doing some public relations work for the Challenge. I found her a good listener as well as a talker, and told her how much I had wanted to compete in the BOC Challenge myself. A week or so later Kim phoned and asked if I would join the race committee as safety officer. It wasn't quite the same as competing, but it would keep me close to the action, so I was delighted to accept.

> *I'd just got the job of working on the race through Hall and Jenkins, who were event managers, and that day at Pier One I hadn't even officially started. I was just there to farewell Nev Gosson. Someone told me to go and jump on this boat if I wanted to watch Nev go out the Heads, so I did. I didn't know whose boat it was. I remember standing on the deck and Ian told me I'd be much more comfortable in the cockpit. I said I was quite comfortable where I was, thank you very much. Later on we became friends, and since it was my job to put together the Sydney race committee, I asked Ian.*
>
> Kim McKay, public relations consultant

As the leaders neared Sydney in December I was at race headquarters every day, tracking positions and listening for radio information on conditions. When the young Frenchman, Philippe Jeantot, was off the Heads leading the fleet, I went out on *Maris* to welcome him. I was terribly impressed with Jeantot, this tough and enigmatic young man who had the best racing yacht money could buy, and the determination to go with it. Philippe stopped short of arrogance, but he was under no illusions about who should be leading the race. He was a quiet achiever, a man who didn't need to beat his own drum. I thought he was a real hero, and so he proved to be by winning the first Challenge in just over 159 days.

As the fleet made its way up the coast, there was plenty of drama. Englishman Desmond Hampton had chartered *Gypsy Moth V* from the widow of Sir Francis Chichester. Now, coming out of Bass Strait, he ran aground on Gabo Island and watched the historic yacht smashed to pieces by the surf. It made us all aware of the dangers involved in this sort of race; dangers which are compounded by exhaustion and unfamiliarity with a coastline.

Jeantot rounded Cape Horn on Valentine's Day, 800 kilometres and three days ahead of the South African Bertie Reed. Now he had an insurmountable lead and crossed the line at Newport on 9 May 1983 in 159 days, two hours and twenty-six minutes.

CHAPTER TWELVE

THE SPIRIT OF SYDNEY

In 1983 Australian sailing got a tremendous boost when Alan Bond's Ben Lexcen-designed *Australia II* came from three-one down to win the America's Cup off Newport, Rhode Island.

Australians who knew absolutely nothing about sailing stood up and cheered that day. It was a famous victory; we had won the unwinnable, whipped the Auld Mug from under the Yanks' noses. For those of us who knew the sport intimately, it was, I suppose, a turning point. Ocean and offshore sailing was no longer considered to be just the province of a rich elite. It was something in which we could all share.

Although he was not a close friend, I'd known skipper John Bertrand for some years and had been team-mates with him at the Admiral's Cup in 1977. Other crew members like Hughie Treharne were good mates. So we shared their joy that day, but it also underlined for a small group of us in Sydney the image problem we faced as single and short-handed (two) sailors. The general public thought we were crazy; a large section of the yachting fraternity thought we were irresponsible.

By this time Kim McKay had agreed to help me, and we were actively seeking sponsorship for my tilt at The BOC Challenge in 1986–7. It was in our interest to do something about the image of solo sailing if we were to gain any advantage from the increased attention being focused on the sport generally. And beyond all that, I was a bit pissed off that we were sailing's poor cousin, struggling to raise a paltry $60,000 when an America's Cup campaign cost millions.

When I thought about it, I'd been a solo sailor in my heart since early 1975, when I'd returned to Papeete and found Benny Hawke with a broken leg. Right then I wanted to sail on alone, but I wasn't prepared for it. As it turned out, I found the best shipmate I could have wanted, but I resolved then and there to be ready to go it alone the next time. And now, a handful of years later, I was ready for the greatest single-handed sail of them all. I wasn't a ratbag, nor were any of the single-handed sailors I knew.

Don McIntyre, a solo sailor from Adelaide who had come to Sydney to work on the first BOC Challenge boats, came up with the idea of starting the Short-handed Sailing Association that allowed commercial sponsorship and signage of boats, unlike the yachting establishment. A group which included sailing writer Rob Mundle, Kim and myself, Don's wife Margie, yacht designer Lock Crowther, sailing coach Alistair Mitchell and yachting identity Pat Toolan, formed a working committee under the patronage of famous solo sailor Robin Knox-Johnston. The first general meeting in December 1983 attracted more than 100 like-minded people, and we made our mission for 1984 the establishment of a fully-sponsored two-handed race.

> *It's funny how single-handed sailing had such a poor image in Australia. I mean, look at how well people like Sir Francis Chichester were received here! But there was a very negative attitude from the yachting establishment here, unlike, say, France, where solo sailing had been embraced. Don McIntyre was the first one to say, 'Let's do something about this.'*
>
> Kim McKay

We started running single and two-handed day races and then lined up sponsorship and slotted a major two-handed race with prize money for January 1985. It was to be run between Sydney and Port Macquarie with the finish line at the sponsor's real estate development, a prestigious waterfront estate called Riana Shores. As soon as it became public there was a lot of interest in the race, with many excellent sailors declaring themselves starters. One of the buzzes around the waterfront was that Tasmanian-born radio tycoon Rod Muir was pulling out all stops to win on his new 53-footer *Doctor Dan.*

I didn't know Muir, but for about a year his name had been mentioned in dispatches as the new big player in ocean sailing. A radio industry veteran who had come up with the winning pop music formula for Sydney's 2SM and a hundred clones around the country in the '70s, he had founded Radio 2MMM at the beginning of the '80s when the first commercial FM licences were granted. Triple M was basically a slightly raunchier version of 2SM, but, led by outrageous breakfast announcer Doug Mulray, the station charged up the ratings, swamping all opposition. By 1984, Rod Muir was seriously rich and totally obsessed with making a mark in sailing.

We had some mutual friends, Muir and I, and I heard around the waterfront that *Doctor Dan* was potentially the fastest boat in the fleet and that Muir had an experienced young bloke from New Zealand crewing with him who could help him beat the pants off all comers.

Preparations for the two-handed race were interrupted briefly by a long-awaited event on my social calendar. My divorce had finally come through and on 11 August 1984 I married Judy Humbley. The wedding was a quiet one at home in Kirribilli, but later Alan Morris hosted a reception for about 150 family and friends at his big house in Waverton.

The Crash and Burners commissioned Jack Earl to paint a wedding gift for us and Jack presented us with a beautiful painting of the nineteenth century cutter *Rover's Bride,* which was the first trading vessel on the Lord Howe Island run. The old man had a wry grin on his face when he gave it to us, thinking, I am sure, that it was appropriate in more ways than one. He had written to me during one of my voyages, 'Stay on the steady side of your wild ways, my roving son.' And by this time, of course, he really was my spirit father. Perhaps the roving days are over, Dad, I thought but didn't say.

In October 1984 I phoned Tommy Stephenson at his Brisbane boat yard and enquired about chartering *Madame Defarge* for the two-handed race. *Madame Defarge* was a light displacement 50-footer from a Ben Lexcen design that Tommy had built for his father-in-law. It was sleek and quick. 'Not interested,' he said. 'But I'll race it with you.' Done!

By now there was a large fleet which included Rob Mundle on his big catamaran, *Eye Witness News*, and several other worthy contenders, but as far as I was concerned, this was a race between the *Doctor* and the *Madame*, between Muir and me. I had no axe to grind with Rod Muir. In fact, although I still hadn't met him, I was pleased to see someone of his wealth and obvious enthusiasm entering the sport, particularly in this event that was so important to us. When I finally met him, Rod Muir did not disappoint.

He was a big, loping fellow, ruggedly handsome I suppose, who mooched around the Cruising Yacht Club in topsiders and baggy shorts, mumbling asides and funny one-liners to his brat pack in a staccato, side-of-mouth delivery that was difficult for the uninitiated to follow. For all that, he was as sharp as a tack, and if you could follow the drift of his conversation for more than a minute, you generally learnt something. I found him immensely likeable, possibly an eccentric genius, certainly a breath of fresh air in sailing. And in for a bloody hiding in the Riana Two to Port race; Tom and I would make sure of that.

Tommy arrived and we spent a week preparing *Madame Defarge*. When the gun went off we were ready. We reached out of the Heads near the front of the fleet, got to the lead by Broken

Bay and stayed there. *Doctor Dan* made up some ground early, but with the northeast sea breeze in, it turned into a tacking duel up the coast with Rob Mundle's big cat, which ended up taking line honours. The going got heavier late in the race and I remember surfing up the river mouth to the finish line. A great race but we were never in serious danger of losing to the *Doctor*.

By the time *Doctor Dan* crossed the line Tommy and I were well and truly in party mode, and that evening we gravitated to Muir's boat to drink his seemingly inexhaustable stock of French champagne. Muir's wife, Kathy, was there, and Kim McKay, along with some other friends, and Muir insisted on playing Trivial Pursuit. Having lost the race he seemed determined to win the game. I think I annoyed him somewhat by refusing to play and calling out the answers.

As the evening progressed the conversation got wilder and wilder, and Muir started enthusing about his sailing plans. At one point he said: 'What we need to do is build a boat called *Spirit of Sydney* and have it represent Sydney in the world's number one single-handed race, the BOC. Yeah, we need Lexcen to design it. We'll get Benny, we'll get the best bloke in Australia to drive the bastard . . .We'll . . .'

He went on and on as the champagne flowed, and I kept thinking, 'He's describing my dream. This is me, this is my boat.' But there were a lot of people in the saloon and we were all drunk. I kept my mouth shut.

> *The BOC race was on my yachting calendar, had been for some time. We never saw it as a 'one guy against the world' type of deal because that was not how the French were playing it. They were winning races like this by using a full team, and that was what I intended to do. I was also looking at my options for selecting a jockey. What I wanted to do was trial two guys and select the one who performed better.*
>
> Rod Muir, former businessman

> *Rod was looking through a book called* Great Yacht Races, *which had a chapter on the BOC Challenge. When he started talking about putting a boat in the next one, I just looked at Ian and thought, 'This is what we've been working at, say something!' But he was right. Keeping quiet was the best tactic.*
>
> Kim McKay

Say what you will about Rod Muir — and a lot of people have — he doesn't let obstacles stand in his way. Tommy and I had whipped his arse in the Port Macquarie race, so his response was to get us on his team. My invitation came a few weeks after the race on the dock at the CYC. I'd heard around the waterfront that Muir was going ahead with his *Spirit of Sydney* plan and that he had hired 'Chas from Tas' to sail her. Chas was Charles Blundell, one of the sailing world's classic characters, a very useful sailor who could fit all his possessions into two sail bags and was in a constant state of readiness to sink a hundred beers at the bar or fly off somewhere to pick up and deliver a rich man's yacht. As a yacht deliverer he was the best in the business. As a single-handed racer, well, I figured I had a few more notches on my belt.

Muir came up to me and said: 'The word around the joint is that you're the bloke who ought to be driving my boat. Why the fuck didn't you say so?'

'Because you wouldn't shut up for long enough!' I replied.

Tommy was next. Muir phoned him and asked him to sail on *Doctor Dan* for the Queensland autumn racing campaign. That involved the Sydney/Mooloolaba race, the Brisbane/Gladstone and Hamilton Island Race Week. While they were away Muir asked him to take charge of his entire boat program, the enormity of which seemed to know no bounds.

> *It started out as the single-handed campaign with Bicky and buying a 40-footer for the Admiral's Cup, but the plan changed along the way. I controlled all aspects of Muir's boat program for twelve months, and that ended up including the building of*

the Spirit of Sydney, *racing* Doctor Dan, *buying* Windward Passage *and getting her back to Sydney, racing the Trans-Pac, two boats in the Sydney/Hobart . . . It was a bloody huge yachting program, and Muir also had fantasies of developing a marina at the Woolloomooloo Finger Wharf. It was just never-ending.*

Tommy Stephenson

Virtually the first thing I did in this job was to take Muir around the world on a two-week buying trip. We were looking for the best 40-footer in the world. I set it up through my boat-building contacts and we went from yard to yard, from designer to designer, all the way through Europe. After a while, Muir said: 'Tommy, I think I want a big boat.' So I gave him the best advice I could, which was if you wanted a maxi, buy a second-hand one first and see how you like it, don't waste millions. The best secondhand maxi in the world was Windward Passage, *and it was sitting in Courtney Ross's boat yard in Clearwater, Florida, with a For Sale sign on it. We flew straight to Florida and Rod walked through the boat, had a couple of beers in the yacht club, then bought it on the spot.*

Windward Passage *was a legendary boat, of course. It had been built on the beach in the Bahamas and was an instant racing success. Probably the only comparable Australian boat would be* Ragamuffin I. *The* Passage *won races from the day it was launched right up to the late 1980s.* Sail Magazine *in America conducted a reader poll to find out what was the most popular yacht of all time, and* Passage *won it hands down. Muir wasn't just buying a boat, he was buying a legend.*

Tommy Stephenson

Someone from the ABC asked me what it felt like to own a big maxi like Windward Passage, *and I told him it wasn't like that. At best I could be described as the 'current owner'.* Windward Passage *has a style all her own, an aura that is just*

> *awesome. Because I own* Doctor Dan, *I joked about renaming her the* Doctor's Windward Passage. *I thought that was pretty funny, but I was ostracised for months for even thinking that! One person can't own the* Windward Passage. *Just to sail her you need twenty-two good mates.*
>
> Rod Muir, 1985

Muir must have spent more than $2 million on his boat program in 1985. He paid $480,000 for *Windward Passage* and more than $800,000 for *Spirit of Sydney.* Then there were salaries, delivery costs, freight, air fares, hotels, you name it. There was no budget as such, the stuff that had to be bought just got bought. Whenever Tommy tried to be financially responsible, Muir's response was always, 'Don't worry about it, mate. We'll find the money.' Apparently he did, because when he wasn't buying boats he was buying radio stations.

While *Spirit of Sydney* was being built and *Windward Passage* was being delivered, Muir had *Doctor Dan* shipped to Los Angeles for the start of the crewed Trans-Pac race to Honolulu. Tommy, Rod and I flew over to crew her. I was navigator, Tommy was sailing master and Muir was skipper. We had a good race, and Judy flew up to Hawaii to be at the finish line. I guess this was the first time that we saw the other side of Rod Muir.

We'd had a celebratory dinner after the race and the conversation turned to *Spirit* and the BOC Challenge. We'd probably all had a few too many drinks, but Muir became very demonstrative about how *Spirit* should be handled. Since the boat was not yet built and Rod had never sailed solo across an ocean, this was all a bit hard to take. Judy, in particular, was incensed at his behaviour. As we walked through the hotel foyer on the way back to our rooms, the argument continued. As the lift arrived Muir turned to Judy and said: 'And anyway, he's just the driver.'

She simmered as we rose through fifteen floors in the crowded lift, then, as we took our leave, she turned, and gave Rod and the entire lift crowd some fairly precise verbal instructions.

Spirit of Sydney was starting to take shape in Cuneo's boat yard in Sydney, modified from plans drawn by Ben Lexcen for Nev Gosson in 1982. We had gone to Lexcen, a friend for whom I had tremendous respect, and asked him to design a BOC-winning boat. To hear Ben and Muir in conversation together was quite remarkable; both brilliant men in their own ways, both lateral thinkers. They would rave at each other for hours, with the conversation dipping and diving and sometimes heading off on several simultaneous tangents.

Ben had some great ideas, but by this time, in the wake of his winged keel success, he had become an industry, and much of the time, after our initial brainstorming sessions, it seemed we were dealing with offsiders who thought they were Ben Lexcen. As well, many decisions were made by what amounted to a committee, and several of the people on that unofficial committee knew next to nothing about sailing.

> *They rushed the design through. As soon as they got the lines plan they started building the hull, but the rest of the boat hadn't even been designed at that point. Ian had his own ideas about what he wanted, but they were so far behind what was happening at the leading edge of racing yacht design. What he wanted was really a dinosaur from the first era of single-handed racing. He'd based his brief on what had won the previous race, but really, what you had to be doing was looking beyond that at what had happened since.* Spirit *was obsolete before it hit the water.*
>
> Tommy Stephenson

> *Stephenson was a hired hand, and part of his job was to advise on the design and construction of the boat. I don't remember him saying one damn thing that was negative. The boat was right for the race, whether it was right for the jockey is another matter. As for* Spirit *being designed by committee, anyone who knew Benny Lexcen would know that he didn't have much time*

for committees. Benny had unfettered control over the boat.
Rod Muir

There were a lot of bad decisions. *Spirit* was an unusual-looking boat, sixty feet of welded aluminium hull, one mast and not that tall by the standards of the day either. I had paid attention to Philippe Jeantot's *Crédit Agricole* during the 1982–3 BOC Challenge and thought it had the right lines for the race, but my own input into the construction of *Spirit* was more broadly based than that. The problems arose when we had design by committee. Some of the features were superbly designed, like the double-sleeved mast with a small crane on top of the spinnaker block, which was Tom's idea. Like the roller headsail system from Germany and the companion mainsail system we developed ourselves. These were designed to enable me to reduce sail very quickly when I needed to, which can mean the difference between blowing out sails and not.

Then there were other things which, to my mind, were just plain stupid, like Benny's twin centreboards. They cost $60,000 each, and when they were installed I discovered that it took 238 pumps to put one down or one up! This meant at least a twenty-minute effort below deck while you were trying to sail the boat. I'd suggested that we build a model with twin centreboards first, but one of the brains trust just looked at me like I was a complete idiot and said: 'Would you have liked to be the guy who said no to the winged keel?'

The detail in the boat was totally over the top. Muir insisted I have a hot water shower, which was totally unnecessary and I never used, and when I said I wanted a comfortable chair at the chart table, the committee got a Reccaro racing car seat. I had to rip it out and put in something sensible.

No real research had been done on what was required. The biggest problem with the whole thing was you had this crazy man Muir who was paying the money saying, 'He's got to be able to have hot showers and it's got to be insulated and he must be comfortable, a three-burner gas stove and so on.' This was all stuff that Ian wasn't interested in. He eats shit when he's sailing anyway. Muir was telling him the things he needed and Muir wouldn't have a fucking clue! To me Muir had seemed like just what the sport needed, vibrant new blood. He was a radical, and that was good. But I didn't know about his faults then.

Tommy Stephenson

Although Rod Muir was underwriting the entire *Spirit* campaign, his grand plan was to raise sponsorship from the Sydney business community. We got a fundraising committee together but we didn't have a hope in hell. Muir didn't exactly hide the fact that he had money, and his champagne lifestyle was enough to put most people off contributing. The response we got so often was, 'Muir doesn't need my money to do this.'

While we waited for *Spirit* to be completed, I started sailing on the *Windward Passage* with Tom and his crew. The highlight of the summer was the Sydney to Hobart, of course, and *Passage* performed very well with an excellent crew, but already I could see that tensions were mounting within the boat program, particularly between Tommy and Rod. Tommy and I had sailed together enough to know our own faults and foibles. We were close, still are, and when Tommy explained the grave reservations he was starting to have about Muir I had to listen. Part of it was to do with lifestyle. Tommy is a quiet, driven sort of guy, fairly conservative in some ways, whereas Rod was totally flamboyant and often outrageous. He fed off the energy of his own party lifestyle, and sometimes people found it hard to keep up. As Tommy said: 'I'm just not prepared to live his bent lifestyle.'

And with Muir it was difficult to be just a personal friend or just a professional acquaintance. There was no clear delineation. If you were on the team, you were on the team, whether it was racing at three in the afternoon or drinking champagne and dancing on tabletops at three in the morning.

Through all this, I managed to have quite a good relationship with Rod. I admired him greatly and actually liked him too, in his saner moments. He was erratic and impulsive and capable of being totally self-centred, but strangely one of the things that gave him the greatest pleasure was making other people's dreams come true. He knew that by sponsoring *Spirit* he was making my dream come true, and he enjoyed immensely my enthusiasm for the task ahead. Another of his assets was his wife Kathy, an anchor for Rod's renegade speedboat approach to life. She was a delightful woman, and perhaps the first person to bring to my attention the fact that sailors like me were environmental vandals. When we sailed we would often hurl our rubbish overboard, silently repeating the yachtsman's code, that two miles out the entire ocean is a rubbish tip. Kathy started out shaking her head in disbelief, then, as she came to know us better, she would lecture us so sternly that we stopped doing it.

We finally took delivery of *Spirit of Sydney* in early 1986 and I started sailing her outside the Heads with some of the *Windward Passage* crew. She handled very differently and took a lot of getting used to, which was unfortunate because I didn't have a lot of time. The twin rudders were very heavy to steer and didn't give you enough feel, and the shallow draught made her easy to heel over. On the other hand, down below she was a work of art. The finish was exquisite and the toys were state of the art. I had three computers, two self-steering systems, satellite navigation gear and a radio phone powerful enough to dial virtually anywhere in the world. All of this was powered by a series of generators, solar panels and a small diesel alternator. *Spirit* was frightening and fascinating at the same time.

I'd only sailed her a couple of times when Muir's mate Michael Chugg, the concert promoter, phoned me. Chuggie had Bob Dylan in town. Would I take him sailing? I'm not much of a groupie, but Bob Dylan . . . well, that was different. The man was an icon. I said to Chuggie: 'If the question is Bob Dylan, the answer, my friend, is blowin' in the wind.'

Dylan turned up at the wharf with an entire platoon of

minders and a very beautiful black woman. Tommy and I and Tom's wife Penny were the crew, and we stood in line for introductions, but the icon and his woman swept past us and sat up front on the cabin top. The minders left and we cast off. Tom and I wondered aloud about the safety of the passengers, but there wasn't much breeze on the Harbour, so we shrugged our shoulders and left them where they were. The plan was to do a quick sweep along under the Harbour Bridge to show off the city, then head up to Store Beach for a picnic lunch and a swim before heading home.

Right in front of the Bridge we hit a sudden and nasty squall, and *Spirit*, as was her shallow-draught habit, heeled right over, almost dunking the boom. Dylan's girlfriend screamed and slid right off the cabin top into the water. While Dylan went into a rather noisy panic — the most animated we'd seen him to this point — we quickly hauled the girl in and took her down below, where we discovered that she'd scraped quite a bit of skin off her leg, opening up some rather painful wounds.

Dylan looked at this mess and announced: 'Okay, that's it. We're outta here. Take us back to the wharf.' The girl was terrific. She looked at Dylan like he was a spoilt little boy, explained that she was fine, that she would be really disappointed if we didn't go ahead with the picnic. The singer took some convincing, but he settled down and seemed to quite enjoy the sail up the Harbour. At Store Beach he went beserk, jumping off the boat into the clear, green water and exploring the deserted beach, swinging from the trees and whooping with delight. Tommy and I looked at each other in amazement. Perhaps he'd never done anything like this in his life.

We were on the campaign countdown. As we put our team together *Spirit* became the talk of the waterfront. Did Muir know what he was doing? Was *Spirit* a winning boat? Was Kiernan a winning sailor? Everyone had an opinion and not all of them informed. There was some scuttlebutt around the bar at CYC, where Muir was something of a fixture. (In fact he'd rented a flat opposite the club.) Sailing is a sport in which money and ego often play too important a role, and that situation breeds instant experts. On a couple of occasions I had more than a quiet word in the ear of the noisiest experts.

The backup team put together for the *Spirit* campaign was quite remarkable, considering that this was a single-handed race, and sometimes it seemed to me that I had been relegated to the role of jockey while the backroom boys made all the decisions. This was not the way you sailed a boat, no matter how high-tech it may have been. At least, it wasn't the way I sailed a boat. Muir made the money decisions, Tommy (increasingly on the outer) made the campaign decisions, Gargles (the sailing master of *Windward Passage*) was my sailing coach, Roger 'Clouds' Badham was the weather coach, radio boffin Bill White was the communications expert and Kim McKay handled our public relations. For me, however, the most important role was that which, in less politically correct times, we used to call 'boat nigger'.

Boat niggers had become 'boat Nigels' or BMWs (boat maintenance workers).

In my view, one of the best of these was Stevie Bakker, a toolmaker by trade and a sailor of great experience who had helped build the innovative mast for *Spirit*. Muir had promised the job to someone else, but I held out for Stevie and got him. Stevie came as a package with his girlfriend Susie Cox, and essentially their job was to be in each port with a full backup service — accommodation and transport organised, provisions sourced and spare parts ready to perform any work that might need to be done on the boat. As a rule, 'boat Nigels' were characters who loved sailing and knew their boats better than any owner. Stevie was no exception, but he also became a good friend and sounding board. He was a quiet man, difficult to know but worth the effort.

Spirit had to be at the starting line in Newport, Rhode Island, by early August, and before then we had to thoroughly trial her in ocean crossings and I had to sail a 2000-mile (3,200 kms) qualifier. I elected to do that on a run I knew well and loved, New

Zealand to Tahiti. So, on a stormy evening in early autumn Stevie Bakker and I sailed out of Sydney into a rising gale.

> *The boat's first real trial was when Ian and Steve Bakker sailed to New Zealand. I heard that stuff was falling off it and there was all kinds of shit, but I had my own problems by then. I just told Ian on the phone that he couldn't take a broken boat to the starting line, to spend what he had to. Muir and I had some ferocious blues and after the Hamilton Island series on* Windward Passage, *I got my marching orders. No explanation, just a Don't Come Monday. The* Spirit of Sydney *campaign was away without me.*
>
> Tommy Stephenson

> *In my opinion,* Spirit *sailed beautifully across the Tasman.*
>
> Steve Bakker, boat maintenance worker

CHAPTER THIRTEEN

AROUND ALONE

Spirit limped into Whangerei with a lot of problems. I had phoned ahead and got Marty Farrand to line up a boatyard and Stevie and I slipped the boat immediately.

This wasn't particularly unusual. Good racing boats are notoriously fragile — look at *oneAustralia*, which sank off San Diego with no warning! And you can't expect to build a 60-footer and just go sailing. You need to see what works and what doesn't, and fine-tune as you go. At Whangerei we did about $30,000 worth of fine-tuning over eight days in the boat yard. Boats like this are similar to Formula One racing cars in that they need constant attention and maintenance after a run around the track. However, I wasn't used to spending other people's money like that and it didn't sit well. I had permission from the manager of the boat program, but Tommy's position was being eroded by the day, and before I left Whangerei, Muir had sacked him. Now I felt I had no allies at headquarters.

I left Stevie with Marty in New Zealand and sailed on alone . . . or so I thought. There were problems almost immediately. Roger Badham, the well-respected weather coach ('Clouds' to his mates), relayed explicit instructions that I should go north, then east, and so on. Weather information was useful, but I had been along this sea road several times before and I took exception to the idea that I needed now to be directed. For Christ's sake, give me the weather predictions and let me plot a course through this sea I know so well! (Weather coaches were later banned in the BOC Challenge because of the unfair advantage they gave competitors who could afford them!)

I started to realise there was a problem when Muir intercepted messages between Clouds and myself. His own message came through loud and clear: 'Do what you're told.' What was I, a monkey in a space capsule? As I sailed on towards the Cook Islands I started to realise just what I had got myself into. I had to pull rank now, or forever be considered the dumb pilot, the crash test dummy. And the fact of the matter was that despite the banks of computers and access to world weather charts, sometimes weather experts predict the weather you've just had. I ignored some of Clouds' initial instructions and continued on the course I felt appropriate, since I was there in the middle of the Pacific and the members of the support team weren't.

Perhaps this could be construed as an old dog refusing to learn new tricks, perhaps a lot of things. I just took it as the responsibility of a skipper at sea, which is to guide your boat to the best of your ability, in the known conditions. Everything the support crew could throw at me was valuable, sometimes invaluable, but it was never the same as being on the boat.

I encountered a gale when I was close to the Cook Islands and *Spirit* slammed and banged along close-hauled. It surged ahead pretty well but the noises worried me a bit. Then the weather got lighter and lighter and I began to notice that the boat was trimming rather strangely. Something was wrong. *Spirit* had a number of vertical watertight bulkheads and two inspection hatches built into them. I opened one of them and discovered it was three-quarters full of water, with the anchor rope sloshing around in it. The weight of the water was pushing the bow down and causing the boat to be out of trim. I'd been through a gale with a lot of water over the deck, so my first thought was that it had leaked in through the stauncheon bolts, but before I started to pump it out I got a torch and went inside to investigate.

I crawled through the bulkhead into the black and flooded compartment and shone the torch around. Straight down I could see a strip of electric blue. I knew what it was immediately and it horrified me. It was the Pacific Ocean. The bow had split.

I hurriedly pumped out the bulkhead and attempted to repair the split, using timber and epoxy. It was difficult work because I had to pre-cure the epoxy on deck in the sun, and then hold the timber and the epoxy over the split while it took hold, but I managed to slow the leak down.

I had completed the 2,000 solo miles needed to qualify for the BOC Challenge and estimated that I was about two days south of Papeete, so I radioed race headquarters and explained that I had a major leak and requested permission to log my position and motor into Tahiti. They agreed to the request and told me to arrange to be met by officials from the Tahiti Yacht Club to sign the log and work out how much fuel I'd used and validate my qualifying voyage.

A day or so later I watched a small flotilla of rubber boats appear on the horizon and head towards *Spirit*. Soon I could make out figures. Stevie was on the lead boat. After the commodore had been through the formalities several of the Tahitians stayed on board to guide me into Papeete. I said to one of them: 'Do you know my friend Rodo Williams?'

He said: 'Oh yes, Rodo went beyond the reef just last week.'

I knew what that meant. It was the Polynesian exit, a death with great dignity. When Polynesian men who value the traditions grow old and frail, they choose the moment to take a canoe and paddle, just before dawn, to the deep water beyond the reef. There they quietly slide overboard, consigning their spirits to the depths. Another mate gone, but there was no time for tears.

Marty Farrand had flown in with Stevie to help sail *Spirit* on to America, so naturally the three of us had a quiet little drink with my French and Tahitian friends. Returning to *Spirit* to sleep it off, I tripped on the gangplank and fell onto the stern rail, breaking two ribs in the process. I was therefore feeling somewhat poorly when we convened in the morning to start the work on the boat. Stevie had organised a couple of cranes to lift *Spirit* out of the water and over the sea wall, but the cranes were dreadful old things, full of rust and with frayed slings. I said to Stevie: 'I haven't got the stomach for this. If they drop the boat, I'm out of the race and we all look stupid.' We sent the cranes back and negotiated the use of another smaller one which was subsequently used to lift the bow out of the water so that the 600-millimetre split could be reinforced and welded.

I flew home to Sydney to try to put my business affairs in order for the long months I'd be away. An accountant friend named Ian Ferrier had put in place a 'scheme of arrangement' with my creditors in the building industry to enable Alfred Builders to continue trading, but these were incredibly difficult times. I had put everything into the *Spirit* campaign and I had been supplied with, arguably, the best of everything. But I wasn't getting paid and this had placed an enormous strain on my business. A dozen years on, I could see the problems of 1974 looming again, but this time I knew how to deal with them.

> *Stevie and I sailed direct to Panama in twenty-eight days. We carried four forty-four-gallon drums of diesel plus a full tank and if we dropped below six knots we hit the power button. We weren't sailing, we were delivering, but I must say I wasn't very impressed with the way* Spirit *handled. We got her to Clearwater, Florida, and pulled her out of the water at Courtney Ross's boat yard.*
>
> Marty Farrand

Spirit remained in Courtney Ross's boat yard — where Rod and Tommy had bought *Windward Passage* a year earlier — for several weeks, undergoing the biggest overhaul of her young life. We put a new bulb on her keel with extra lead ballast to stabilise her. The work was intense in the dreadful sticky heat of a Florida Gulf Coast summer, and as the yard bills mounted, I found myself concerned about spending other people's money again. I guess that concern was well known at the yard. One day a message came over the PA: 'A phone call for Ian Kiernan. Trish from Triple M in Australia calling.' Trish handled the accounts for the *Spirit* campaign. I knew I was in trouble.

I got to the phone. 'Had you shittin' yourself there, eh buddy!' It was Courtney Ross, the bastard.

> *I think that the modifications made in Florida actually hampered the boat's performance, but that wasn't for me to say.*
>
> Steve Bakker

Stevie, Susie and I sailed down the Gulf to the Florida Keys, then headed up the coast to Newport, making good time. One night, with a good following wind, we cracked sixteen knots and roared with delight at this new top speed. Newport, Rhode Island, is a beautiful little city, rich in maritime history and alive with the spirit of sailing. This was particularly so with the BOC fleet in town and excitement mounting as the race drew closer. We rented the bottom half of a little clapboard house near the waterfront. One of the French teams had the top half.

One by one the competitors sailed in and tied up at the Goat Island marina. In Class 1 (50 to 60 feet length) there were 1983 winner and hot favourite Philippe Jeantot of France in a new boat, *Crédit Agricole III*; his two compatriots Titouan Lamazou and Jean Yves Terlain in *Ecureuil d'Aquitaine* and *UAP* respectively; French-American Guy Bernadin in *Biscuits Lu*; New Zealander Richard McBride in *Neptune's Express*; fellow-Aussie John Biddlecombe in *ACI Crusader*; American David White (the race's founder) in *Legend Securities*; Warren Luhrs, also from the US, in *Thursday's Child*; and South Africans John Martin and Bertie Reed in *Tuna Marine* and *Stabilo Boss*. In Class II (40 to 50 feet length), five Americans, Mike Plant in *Airco Distributor*, Dick Cross in

Airforce, Hal Roth in *American Flag*, Mark Schrader in *Lone Star* and Mac Smith in *Quailo*; Czech-American Richard Konkolski in *Declaration of Independence*; Canadian John Hughes in *Joseph Young*; Englishman Harry Mitchell in *Double Cross*; two Finns, Harry Harkimo and Pentti Salmi, in *Belmont Finland* and *Colt By Rettig*; Brazilian Eduardo Louro de Almeida in *Miss Global*; Takao Shimada of Japan in *Madonna*; and Frenchman Jacques De Roux in *Skoiern V*.

My fellow-Australian John Biddlecombe had started his campaign in sensational fashion when he'd wrecked the original *Crusader* on a reef in Tonga while doing his qualifier. (He'd overslept his watch and not slowed the boat down approaching his landfall. Obviously there was no tutor like Jack Earl in his life.) Amazingly, his sponsors had footed the bill for another *Crusader* — a boat built for entrant Claire Marty who had to withdraw due to lack of funds — and Biddlecombe was here at the starting line. He was a short, muscular fellow who looked as if he pumped iron every day. I have to confess, I didn't understand the man at all. Nor did the French, and he worried them. They saw him as the kind of maverick competitor who might break all the rules and somehow win. One day Biddlecombe came up to me on the dock and asked: 'Do you climb the mast every day?'

It was a very strange question. I climb the mast only in a dire emergency and regard anyone who does it for fun during a race as a fool. However, I had just climbed *Spirit*'s mast in Newport, using a new French system, which I told Biddlecombe about. He said: 'Oh no, I mean hand over hand, with no bosun's chair.' I scratched my head and walked away.

Mark Schrader, who was later to become BOC's race director for the 1990–1 and 1994–5 Challenges, obtained some support from the Center for Marine Education. While we were preparing our boats at Goat Island he told me that, to draw attention to the Center's support, he was planning to store all his plastic debris on board the boat, rather than throw it overboard. It would be interesting to see how much garbage one sailor could have hurled into the ocean. Remembering Kathy Muir's distaste for the yachties' waste disposal system, I told Schrader I thought this was a great idea, and that I would collect my garbage too.

There were twenty-five boats in the race and I thought *Spirit* stacked up reasonably well. There were perhaps more question marks over her performance than there were for the sleek, spartan speed machines of the French, and the centreboard cases on deck attracted some quizzical stares, but she looked the part. I knew that together we would be competitive, and I thought we might just win.

The *Spirit* camp grew larger each day. Rod and Kathy Muir arrived and moved into another rented cottage. Then came Judy, Kim McKay to handle the publicity, Bill White to set up the communications base and several others to just enjoy the party. For me, however, there was no party, just added pressure. I seemed to spend at least part of every day being told what to do, and this did not sit well with me. Matters came to a head when Muir called a team meeting at his house. I was presented with a photograph of Bob Hawke shaking my hand as he farewelled *Spirit* in Sydney. Then they got down to business. I was put under the microscope and every aspect of my handling of *Spirit* came under attack. Again, it was being made clear to me by Muir and everyone else that I was no more than the jockey. Judy and I got up and left.

On the way back to our house she burst into tears. She said: 'I've got my credit cards, let's just get a cab to the airport, buy two tickets and go home.' It was tempting, but I couldn't walk away from the race. It was too important to me. No matter how much I had to put up with, I was going to sail *Spirit of Sydney* around the world.

> *Rod was most unpleasant that night. He laid down the law about Ian making phone calls. He wasn't to phone me, he was only to phone Roger Badham every day. He was the jockey and his job was to sail the boat. It went on and on. It was an appalling way to treat Ian so close to the start of the race.*
>
> Judy Kiernan

A lot of money went into that campaign and as chief executive of Triple M, I was responsible for the expenditure. I had to make sure that money was used the right way and I believed — still believe — that operating as a team was the way to win the race. We trialled the boat for days on end in Newport, and most days Ian wasn't even on board! He wasn't playing the team game. I flew his wife in before the race started, hoping that would have a calming effect, but that didn't work too well.

Rod Muir

The problem was that we had the right boat in the wrong race. The prospectus for the race had concentrated on its being a round-world affair, whereas it was really four short legs. So, in a sense, it had been overbuilt. I think both Ian and Spirit *would have been highly competitive in a non-stop race, but when we got to Newport and saw what the French had, well, we just started to realise there hadn't been enough research done.* Spirit *was the lady of the fleet, sitting there amongst French boats that were just racing mongrels. Naturally, Rod started to ask some questions. Like, how come we didn't know about all this?*

Steve Bakker

The next day the diesel power generator blew up. We were thirty-six hours away from the start. Muir moved heaven and earth and got another motor flown in from Texas. Then Stevie Bakker and a team of mechanics had to work through the night to get it installed in time. Nuts were still being tightened as I was towed to the line.

At dawn on the morning of the race Ian and Warren Luhrs were to be interviewed by Bryant Gumbel of the Today Show *on NBC. I'd briefed the program's researchers on how funny and witty Ian was, and gave them some story leads that would bring out the best in him. Obviously that morning his mind*

> *was completely elsewhere. The fighting with Rod, the pressure of the race . . . it had got to him. He was terrible. It was the worst media performance of his life.*
>
> Kim McKay

On the morning of the race I walked along the dock to *Crédit Agricole III* and said to Philippe Jeantot: 'Good luck, mate.' He looked rather surprised but I meant it, even though he and Biddlecombe were, for different reasons, the people I most wanted to beat.

It was a beautiful Newport morning with several thousand people lining the foreshore and a huge spectator fleet on the bay. Among the race fleet there was a lot of tension which resulted in a few mishaps. *Tuna Marine* hooked the mast of a spectator boat with its leeward runners, dragging it sideways and dumping the crew in the water. *Thursday's Child* had a collision and had to go back into port. And in my excitement I broke the start and had to go back through the fleet and cross the line again.

Despite this, I soon recovered and headed out into the Atlantic near the front of the fleet. After so many months of preparation, after so much tension, the race was on. Jack's words came back to me as I plotted my southeast course: 'Cool your excitement, old friend.' I tried to.

Bill White was to set up communications points along the coast as I sailed down towards the equator, and Stevie had been assigned as his driver. (This was something like putting Idi Amin and Queen Elizabeth in the same phone box!) They would park their radio caravan on headlands and wait for me to tune in for our 'scheds'. The number of scheds each day was absolutely ridiculous and, as I sailed further away from the coast, transmission became more difficult. I felt it was more important to sail the boat than be tied to the radio, so I started to miss scheds and after a while, stopped doing them at all. This was becoming an increasing frustration for Bill White, who took it out on Stevie, who, in turn, solved the situation by abandoning him.

Meanwhile, my radio phone enabled me to stay in touch with those I wanted to. I phoned Judy, my weather coach and even a few old mates in Sydney. Anyone but Muir. I was still seething over his treatment of me in Newport. One of my mates ran into Rod at the CYC bar. 'Had a long chat to Bicky the other evening. Says he's going well but could do with a bit more breeze.' Muir was furious. Here he was spending millions on this campaign and he had to be told in the bar how his 'jockey' was doing! This was a smack in the face to Rod, but it was not something I'd planned. It never occurred to me that these people would meet. Nevertheless, considering the events that were to follow, I would have been better advised to use the phone more selectively.

> *The phone bill was ridiculous. He was out there having a picnic and I was footing the bill! Quite frankly, the fact that he was on the phone for so much of the time was starting to make us believe that the boat wasn't being sailed to its capacity. It made me mad because the land-based support team was working so hard. Jesus, when he was labouring in the doldrums we even dragged Bob Hawke out of a Cabinet meeting to send him a cheerio call!*
>
> Rod Muir

I caught the tail end of a couple of late season hurricanes to help me over the Gulf Stream, and registered breeze up to twenty-eight knots. Then, off to the southeast of the Caribbean, I got into light air and stayed in it. The doldrums, a windless band that stretches across the Atlantic just north of the equator. Stinking heat, water like oil, the soles of your feet burnt from walking on the hot deck, the incessant flapping and cracking of sail as you roll in the swell. Kathy Muir had convinced Rod that *Spirit* had to have the best sails money could buy, so he'd spent an extra $50,000 on ultra-lightweights for conditions like this. And now they slapped and cracked so violently in the swell that I wondered if I might blow them out without a breath of wind!

Occasionally a line squall came through, and one caught me napping, its twenty-five knot breeze blowing out my light jib. I had to sit on deck in the heat and sew it together again just to get going. The frustrating conditions continued. I was quite close to Jean Yves Terlain on *UAP*, but he managed to pick up a light breeze and break through. *Spirit* just wouldn't move. I racked up a mere thirty kilometres day after day. Clouds, my weather coach, kept telling me to 'head south, don't cut the corner'. I kept heading south as the rest of the fleet sailed past me. The only sleep I got was a catnap in the cockpit because I couldn't afford to miss the slightest change in the weather. The days developed a pattern; at sunup I'd run around the deck and check on everything, then prepare some muesli and fruit for breakfast, then go through my work list, navigating, repairing and so on. After a snack for lunch I'd try to get another hour's sleep, then work at the chart table and on the radio. The satellite communication system meant that I had good contact with the other competitors and, although hearing where they were was sometimes depressing, it developed the camaraderie between us and made us all feel anything but isolated. People have often asked me how I handle the loneliness of a solo ocean crossing. I never think about being lonely, particularly in a race fleet where every moment of the day is taken up with extracting maximum performance from the boat, but certainly there is a different feeling that you have to get used to. You live with the sounds of the boat and the sounds of the sea, and it sometimes feels like there's a strange canopy over you, or an aura surrounding you. Not unpleasant, but strange.

I'd provisioned with quite conventional food, but during that first leg the fridge blew up and I had to throw most of the food overboard. I went back to a diet of fruit, rice and canned sardines.

Finally the light conditions were behind me, and I picked up a sou'easter and *Spirit* started to roar as I celebrated my forty-sixth birthday. I had a Super 8 movie camera on board so I set it up and filmed myself waving a sign, '46 today'.

What blessed relief to be doing ten and twelve knots on a sea of white caps again! I flew across the South Atlantic and saw magnificent Table Mountain pop up on the horizon. Then more frustration. The breeze dropped as I closed with Cape Town making very little headway against the north-flowing Benguela current. After two difficult days I made it into Cape Town with my half-ounce spinnaker up, jibing and cursing as I closed on the finish line. The first battle was over. Newport to Cape Town in fifty-one days. It was not a great result, but I was there and I wasn't the last boat in Class 1 to cross the line. (Biddlecombe was still in Bermuda, having a new keel built!)

Stevie and Susie had set up our digs in Cape Town and Kim McKay had flown in to handle the public relations. I went straight to the apartment and shaved and showered off the salt of the Atlantic, then joined the team for a late drink. I asked Stevie: 'How's Muir? Has he said anything about my position in the fleet?' I hadn't spoken to the campaign boss for fifty-one days and I still didn't feel I could without turning it into a screaming match. There was no answer. Obviously we had some problems, but I was too exhausted to care. It could wait until I'd rested.

We convened again at breakfast. Stevie said he'd spoken to Muir. There was a long silence again. I said, 'Well?'

Stevie said: 'It's not good, mate. Muir says he's had it. The boat's out of the race. He'll fly the *Windward Passage* guys over and see if they can set a few ocean records taking her home.'

'Over my fucking dead body.' I got up and walked out of the cafe. I needed some air.

> *Ian took it very personally, which I could understand, but I could also understand where Rod was coming from. There are two ways you can enter the BOC. One is the French way, where you put everything into the boat and the support crew, and you aim to win it. The other way is what I'd call the British way, where the challenge is to circumnavigate the world and it doesn't matter where you finish, you just enjoy the camaraderie and the character-building aspects of the race.* Spirit *had been*

built to win and had a huge backup crew, but we weren't pegging it with the French. There was nothing wrong with saying, we can't win a sprint with a distance runner, we're out.

Steve Bakker

I had a responsibility to the company which had invested a lot of money in Spirit of Sydney. *That was all there was to it. I felt we could achieve the best result for everybody by withdrawing from the race and sending a crew over to see if we could set a new record from Cape Town to Sydney.*

Rod Muir

I had a dual role at this stage. I was handling the publicity for the race, but I was also working for Rod Muir, so I was in contact with him frequently before I went to Cape Town. Rod had a chart on his wall where he plotted the positions of the Class I boats, and around it he had a collage of photos of the race start, all the boats and skippers. What I found remarkable was the fact that there wasn't a single photo of Ian up there. It might have been just an accident, but my own view was that in Rod's mind it was the boat, not the sailor that was the hero. And in this case he seemed to think that the sailor wasn't even doing his job! I remember on one occasion Kathy Muir was there and she said: 'Kim, I just don't understand it. If this is Ian's dream, why is he doing it so badly?' What could I say? It was sailing, things don't always go to plan.

Rod gave me a ticket to Cape Town to represent the team. Ian slowed down, so I was there a week before he arrived, and during that week both Stevie and I spoke to Rod several times. I had a screaming match with him on my birthday, October 20. I remember him saying he was going to pull the boat out of the race. I told him he couldn't do that. He was adamant. He said: 'You're working for me, you'll do what I say.'

Kim McKay

CHAPTER FOURTEEN

DOG DAYS AND SOUTHERN OCEAN NIGHTS

It took me several hours to come to terms with what Muir had said, and I decided I couldn't accept it. I said to Stevie: 'I'm no quitter, mate. Possession is nine-tenths of the law and I've got the boat. If he wants to play hardball I'll lock it up.'

And I meant it. Since the campaign began I had had a big sword hanging over my head. I would wake up some nights in a cold sweat, wondering aloud: 'What if I don't finish?' History doesn't remember the boats that don't finish. The greatest challenge is to win, but the next greatest is to finish. There was no way I was going to be robbed of that.

Kim decided that she would phone Muir again and try to mediate in the dispute.

> *He seemed more conciliatory this time. He was prepared to keep the boat in the race, but not at his expense. He paid for slipping and Stevie worked out a deal with him where he also continued to pay wages for him and Susie, which made sense, since their job was to look after his boat, his investment. But it costs a lot of money to sail a 60-footer around the world, and the tap was literally turned off.*
>
> Kim McKay

She came back to me and said, 'He's pissed off but he's calmed down a bit. He says he'll pay to have the boat slipped, but if you want to keep going, you're on your own. That's as far as he'll go.'

I had about a hundred bucks in my pocket at the time, there was a lot of work to be done on *Spirit* and many more costs to be covered if I was to stay in the race. The problems that this withdrawal of sponsorship created were immense, but I felt as if a huge weight had been taken from my shoulders. If, somehow, I could keep *Spirit* going, I would effectively be chartering her. It would be my boat, my race, my decisions. But first I had to get the money.

Kim leapt into the rescue operation in Cape Town, sending out feelers for sponsorship while keeping my row with Muir out of the media. Meanwhile, I phoned home to see what could be done. My family and Judy's, and my friends, like my old rugby mate Hans Lamens, who organised a substantial whip-around, were magnificent. If I needed money they would find it somewhere. Then CIG (Commonwealth Industrial Gases), BOC's Australian affiliate, weighed in with $10,000, a fantastic gesture. I figured I needed about $30,000 to stay in the race, and by the end of my first week in Cape Town, I had it. I was humbled by the show of support for me from so many people, and I have never forgotten it.

Once I had made my decision to continue the race without Rod Muir's sponsorship, I quickly revived mentally and physically, putting the dreadful first leg behind me. I had had too many things on my mind, I had sailed the boat badly on the wrong side of the course and I had been dogged by bad luck. But that was over now. Cape Town to Sydney was a whole different ball game. I was no longer the poor cousin, sailing the boat and feeling embarrassed and ashamed every time something went wrong. Now I was in control and I firmly believed that I could create my own good luck and turn the race around.

The other competitors were extremely supportive. One day Philippe Jeantot came along to *Spirit* at the Royal Cape Yacht Club and said, 'Ian, can you come with me for a minute?' I followed him to one of the other docks and the whole French team was there, along with South African John Martin, who had won the first leg on *Tuna Marine.* Philippe wished us all well in the second leg and then presented each of us with a full set of his company's oilskins and thermal gear. This was an extremely generous gesture — the gear was worth thousands — and very welcome. At one stage it looked like I'd be sailing the Southern Ocean wearing my old woollen jumper, but then Judy had sent a full set of thermals, and now I had two. I don't know why Philippe had chosen to include me in his generosity — at that stage you wouldn't have picked me as one of the leaders of the fleet — but I thought back to the day the race started and his surprise that I should wish him luck. Maybe good karma was starting to kick in.

At the yacht club I met a friendly bloke called Ian Hershon who invited me to spend some of the time before the race restarted at his home at the foot of Table Mountain. It was quite common for BOC competitors to be invited into the homes of their hosts and a great way to experience the country, but in this case I declined because it meant hiring a car and commuting to the boat every day. But Hershon persisted and eventually I agreed, hired an old VW and moved into his guest quarters. The Hershons were white, well-off and probably politically conservative. At their beautiful home they had a black African 'houseboy', a real gentleman in his late sixties. At first I found this situation rather difficult. I didn't know how to handle the fact that this man was in a subservient position because of his colour, and yet clearly very happy about it and well-loved by the Hershon family.

As I got to know Albert I became a little easier with his position. I'd come home from working on the boat in the evenings and go straight to my room to shower. Albert would knock on the door and appear with a drinks tray. 'Scotch and soda, Mr Ian?'

'Albert, you're a bloody mind reader!' He fell about laughing every time I said this.

I had been interested in low-cost housing since Joint Venture 500, when we had become involved with Aboriginal housing in Redfern, and told Ian Hershon that I would love to visit a black African community. 'Well, why don't we ask Albert to arrange it?'

One Sunday morning Albert took me into a village on the outskirts of Cape Town where there had been recent rioting. I was amazed at what I saw. The villagers weren't interested in housing supplied by the white man's government. They had a great pride in what they had created for themselves out of materials rejected by white society. What looked like a shanty town at first glimpse was in fact a clean, innovative streetscape. The villagers were down on hands and knees scrubbing their bare board floors, while those who had glass in their windows were cleaning them. It wasn't uniformly clean and tidy, but the majority of the inhabitants were really trying to make do with what little they had. I felt privileged to be among these proud people and I thought it a pity that so few outsiders ever got to see this side of life in South Africa.

Hershon, who was Jewish, said to me one day: 'Ian, why don't we get the rabbi to bless the boat?' I thought this was an excellent idea. Anything that might help! So the rabbi came down to the dock and we had a little ceremony and drank some Passover wine. Blessed, rested and provisioned, I was raring to go.

On the day of the start Hershon had a chopper out on the bay and reported to me that there was a good forty knots of breeze at South Point, whereas at the starting line it was dead light. Bearing this in mind, I started with my Number 3 jib on the furler and when the gale came in just on nightfall I was ready for it and snugged the boat down. In the morning I was leading the fleet as we roared around the Cape of Good Hope.

I pushed south, but then made a tactical error in cutting the corner and heading southeast across the Agulhas current, while Jeantot and Titouan struck dead south. I was not alone and on the second morning of the race I saw Bertie Reed on *Stabilo Boss* and Terlain on *UAP* and a big green front moving quickly towards us. It was just a matter of sitting there and waiting until it hit. Then, whoosh! We were off, roaring into the South Indian Ocean.

The shortest route between Cape Town and Sydney is the 'great circle', heading far south into the beginning of the Antarctic ice drifts. But these are dangerous waters and most competitors chose to go only as far south as necessary to pick up the near-continuous low pressure systems which push westerly gales across the ocean.

I began to feel very comfortable with *Spirit* in the heavier conditions. Clouds, who was now working with me from his temporary base in Fremantle where he was part of the America's Cup defence, was calling the weather shots well, and the boat was also sailing well. The French were all at the front of the fleet and I took to joining in their evening radio chats, often talking to Guy Bernadin and Jacques De Roux, who had won Class 2 into Cape Town. Jacques was perhaps my favourite of all the Frenchmen, quiet, polite and an exceptional sailor.

During our evening chats we'd compare conditions and times. One day I excitedly reported that I was roaring along at a then personal best of eighteen knots. Philippe Jeantot was also on the radio. I said: 'Philippe, how are you?'

'Oh, not too bad, thank you, Ian,' he replied. 'I am sitting on twenty-five knots.' This didn't take the wind out of my sails at all, it just inspired me to push even harder.

As we moved into the Southern Ocean the gales intensified and I found my personal best speeds increasing. I broke through the twenty-knot barrier and kept going! This was fantastic sailing and *Spirit* was up to the task, revelling in the conditions while I worked the boat from the cockpit in the wild seas and marvelled at the bird and sea life. I began to move through the fleet, passing Guy Bernadin in *Biscuits Lu* — which was quite an achievement because this was a highly competitive and beautiful boat — and then Jacques De Roux, who had suffered a couple of horrendous knockdowns.

We were all in unpredictable, dangerous conditions where a knockdown was a constant possibility. The winds were mostly from the west, but they could switch quickly to either southwest or northwest, and they blew at unbelievable strengths, creating huge following seas with big grey combers that broke for half a dozen metres down the face. This was okay. I could handle this. A friend who had sailed the Southern Ocean had lent me his log so that I could study the conditions he encountered, so I was psychologically prepared, at least. And I knew from my surfing days at Bondi that you rarely surfed straight downhill: the secret was to corner and keep the bow out of the trough. Often, as you emerged from the trough, you would feel the full power of the wind hit the boat, almost as if the boat was pawing the ocean surface before it burst through a cloud of white foam to roar down the next savage waterfall.

The real danger in these conditions came when a new front came through, bringing with it changing sea conditions, often from a different direction. This would create huge, cross-breaking waves, and if one hit you on the beam you could quite easily be knocked down, mast in the water.

I was in the cockpit one day, strapped in and battling just these conditions when a monster wave hit, knocking the boat down with my seventy-foot rig pointing downhill into the huge trough. It was a very scary moment, with the air filled with spray, making it difficult to tell where the sea ended and the sky started. One night I was hit by another huge cross sea that actually jibed me. Suddenly the spinnaker pole was in the water and the water ballast was pulling the boat down. The boom was pointing straight up at the black sky! Another time I was trying to get some rest in my bunk to combat near-total exhaustion, when a big wave hit the boat and spun it. The impact lifted me off the bunk and sent me flying through the cabin, hitting the control panel with tremendous force. Fortunately nothing was broken apart from the panel. And I didn't even see the wave!

I certainly knew fear on these occasions, but I managed to control it. I knew that if you panicked you were useless, but I still felt that hollow in the stomach often enough. And I automatically considered my fate every time I had to walk the deck in a gale. It wasn't feasible to be clipped onto the boat at all times, so it became a matter of one hand for the boat, one for yourself.

As the fleet leaders neared the Great Australian Bight some of the other competitors began to develop problems. John Martin on *Tuna Marine* had auto-pilot problems and was steering by hand, sitting around the clock in a bitterly cold cockpit. Bertie Reed had problems too and was forced to put into Albany, Western Australia. Even Philippe Jeantot was having difficulties, and while he was pinned down by a gale coming into Bass Strait, I was able to make inroads into his lead on me. I always had mixed feelings when other competitors ran into trouble. There was no great joy in hearing of their problems: in fact, I felt genuinely sorry for anyone who was having a bad time out there. But it was a race, and you took your opportunities where you found them.

I had a dream-run through Bass Strait, hammering away at the big yellow boat, *UAP*, which I had in my sights now. The sight of the Australian mainland lifted my spirits enormously, and *Spirit* ripped along the coast. Even when the wind headed me, I didn't mind. I short-tacked the coast, knocking off the kilometres, loving the views of that wild place on every onshore tack. At Point Perpendicular a new front came through and, with the breeze behind me again, I put a spinnaker on and ran to the finish line.

As I approached the heads of Sydney harbour the camera crew choppers came in overhead and the spectator fleet started to appear. One of the first boats I spotted was the *Triple M* tender, with Muir's Triple M team jumping up and down and waving. Obviously something had changed in my relationship with my sponsor. A couple of radical jibes and I could see South Head off the port bow. I was going to reach the line with my spinnaker on or die in the attempt. I saw *Windward Passage* at the Heads and I made for the line, over-trimmed to hell.

Just as I reached it, *Spirit* did one of her huge broaches, keel out of the water, with people gasping and screaming in the spectator fleet. I was home, fourth boat into Sydney!

Suddenly people were clambering on board, slapping me on the back, pouring Dom Perignon. I hugged Judy and the girls. Muir was there, all smiles and champagne bubbles. He turned to my mother and said: 'The boy's earned his spurs this time.'

> *The wind was blowing so hard on the second leg you couldn't have slowed the boat down if you tried, but at least this time he'd put a couple of sails up! I think the fact that we'd given him a bit of a shakeup in Cape Town had a lot to do with that. But Ian sailed well, he came of age as a solo sailor that leg, and he was certainly the personality of the race. Most of the other solo sailors had had personality bypasses.*
>
> Rod Muir

> *We went out very early in the morning and motored down the coast looking for him, so we could alert the media. We found him and he was just broaching time after time. He seemed almost out of control. I think he was just terribly excited because he'd done so well. He must have broached three or four times between Bondi and the Heads, and the last one, in front of all the media boats, was just extraordinary. His keel came out of the water and people were screaming. The boat seemed to skim across the water towards the rocks at South Reef. Stevie Bakker leapt aboard off the tender and cut the spinnaker free with a knife. It was a wild way to finish.*
>
> Kim McKay

The Crash and Burners had chartered their own boat and were togged out in their green club blazers with yellow piping and heraldic emblems. They'd been out since early morning and were drunk as skunks as they pulled alongside *Spirit.* Long after the race, people would ask me about the crazy scenes that day. 'Who were those drunken officials in the green blazers?'

I was taken under tow to Birkenhead Point and fleet leader Philippe Jeantot was there to meet me. At the press conference I was asked what the hell I had been trying to do at the Heads. 'I didn't want to finish like a wimp,' I replied. 'As my friend the Bilge Rat would say, I think I must have got my ambitions mixed up with my capabilities,' I added.

After the formalities I had a talk with Rod Muir. It was a slightly awkward meeting, but basically the message was that the sponsorship was back on. That was a huge relief to me and I was happy to put the Cape Town problems behind us. To this day I'll never understand why Muir decided to dump me after one bad leg, but he was man enough to admit he'd been wrong about my ability to sail *Spirit.* 'What do you want to do with the boat?' he asked.

'I want to pull those bloody centreboards out and get them out of my life!'

Muir nodded his approval. Stevie Bakker and I took *Spirit* to the CYC and got a crane in place to lift the boards out, like pulling teeth. They seemed strangely reluctant to move, and when we finally got them out we discovered why. With waves washing over the deck they'd filled with water. I'd been dragging a quarter-tonne of water around the world in each of them.

The morning after I finished I was at race headquarters and the race director, Robin Knox-Johnston, called me aside. 'We've got a problem with Jacques De Roux. He's sailing away from the mark at a very low speed, quite erratic and he's not in bad weather. Something is very wrong.'

Jacques had been through hell for a couple of weeks. He'd been rolled a couple of times in the Southern Ocean, he'd bent his steering gear and been hit in the back of the head by his boom. The boat had become a handful and he'd apparently been very depressed as he slipped back in the fleet. After the bold showing the French had made into Cape Town, their sponsors had gone crazy with publicity at home and Jacques felt great pressure on him to perform well. He had talked about putting in to Melbourne to fix his steering, but Guy Bernadin had talked him into pressing on. Now something was seriously wrong.

Jacques knew the dangers. In the 1982–3 Challenge he had been dismasted in the Southern Ocean and forced out of the race when the spar punched a hole in his boat. English competitor Richard Broadhead had turned around and gone back to pluck him off the deck before the boat sank. Now the race committee sent out a plane to locate the boat off Green Cape and buzz its skipper. The plane made several passes and saw no one. It was possible that Jacques was badly injured and couldn't get up onto the deck, so a merchant ship in the area sent out a tender to investigate. The rescuers found dishes in the sink and no sign of Jacques De Roux.

I was standing in the race office when the news came through. None of us could speak. Tragedies like this deeply affect all competitors, because you know that it is simply the luck of the draw. I had squared up to the dangers before leaving Sydney to join the race start. I had updated my will, put my affairs in order as well as I could. I made sure *Maris* was tidy down below because I didn't want people coming on board and saying, 'God, he was a grubby bastard.' I pulled a big white cover over the boat and I remember looking back at her from the dock and thinking it looked a bit like a shroud, but death at sea wasn't something I dwelled on.

The memorial service was held at the South Head chapel at HMAS *Watson* up on the cliff-top overlooking the Tasman Sea. Philippe Jeantot spoke first in English and then in French, remembering Jacques as a fine sailor, true friend and honourable Frenchman. We all felt his loss, and we all had spent a moment or two considering our own fates.

The Challenge began again at the end of January. For a couple of weeks of the layover I had harboured mixed feelings about this, as I suppose any sailor in his home port would. Although I tried not to, it was easy to get sucked back into my working life, becoming involved with the day-to-day problems. Of course this was the last thing I needed to do, as Jack Earl constantly reminded me. 'Plenty of food, plenty of rest,' he counselled. 'The challenge is still ahead of you.' And it was. At 56°30" South, Cape Horn was the greatest obstacle in the round-world course. Conditions there were never easy, and they could be the most hostile in the world.

Psychologically, the other hurdle for me was the fact that *Spirit* had already sailed around the world once. We had left from Sydney, turned east and now we were back. It was difficult to comprehend that she would have to go around again before we were finished.

So off we went again, roaring down the coast. It was Triple M's *Spirit of Sydney* again, but with a lot clearer understanding about who was skippering her. The summer breeze was behind us and we set spinnakers and made haste across the Tasman. I had never sailed south of New Zealand before and was looking forward to exploring sub-Antarctic waters. I didn't have long to wait. In six days we were off the southern tip of the South Island, clawing a course around the aptly-named Snares, before heading deep into the Great Southern Ocean.

Before I left Sydney Jack had presented me with a card to which he'd attached a bottle of fine cognac and a few metal marine artefacts. The card read: 'Ian, when you are in the storm-swept region of Cape Horn, would you please cast over the side this galvanised iron clew-ring and thimble from *Kathleen Gillett's* jib, as a token of respect and admiration to all those sailors who have weathered this mightiest of all capes, and to those too, who have succumbed to its treachery.'

Now, as I steered south, I received word from Judy that Kath Earl, the former Kathleen Gillett, had passed away. I recalled our embrace at the Earl compound when I left to rejoin the race. She was not well then, and she had given me a special hug and a long, deep, loving look. She was saying goodbye in her own way. I loved that woman, as did all who knew her, and I shed a tear or two as I composed a long telegram to Jack. Of all the emotions you deal with alone at sea, grief is perhaps the most difficult. Despite the fact that you might tell people you want to be alone, you don't want to be that alone.

I had been expecting the wild and bitter conditions of the Southern Ocean, but a high-pressure system ridged out to the east of New Zealand and reached right down into sub-Antarctic waters. The water was tropic blue and, even down into the 50° latitudes, I was still tipping a bucket of salt water over me on deck to wash. And along with the balmy weather came light conditions that were almost as bad as the doldrums.

Kath's death may have had something to do with it, but my spirits began to sag, just as they had as I neared the equatorial doldrums in the first leg. I tried to stay in a positive frame of mind but it was very difficult, particularly when the fleet started to pass me by again. Jean-Luc Van Den Heede in the 50-foot *Let's Go* was one who sailed past when I was at my lowest. I was tempted to strike further south in the hope of heavier weather, but Roger Badham advised against this, warning that I was likely to find myself under a low-pressure cell and battling fierce headwinds.

Finally, some breeze, but nothing like the gales we'd experienced in the South Indian Ocean. Still, beggars could not be choosers. *Spirit* and I jumped on what we were given and at last made some ground on Cape Horn. As I approached the South American coast the weather turned frigid. There was snow on the deck and even in my full thermals I suffered in the cockpit. But the discomfort paled into insignificance beside my excitement at rounding the Horn. This was a milestone in the life of a sailor, an experience steeped in the history and mythology of sailing. More people have been in space than have sailed solo around Cape Horn, and yet some of those who have made it have been in the flimsiest of vessels. It was an exclusive club and joining it had long been my dream.

Cape Horn lies at the tip of the South American continent,

Right: Arriving home after the Trans-Pacific, 1980.

Below: (From left) Nat Thompson, me, Matthew Retmock and Gary Payten (Toss) crewing on board *Maris* and enjoying a week's adventure and fishing, Elizabeth Reef, 1982.

Left: *Maris* departs Lord Howe Island bound for Sydney, January 1983. 'Toss' crewed back with me.

Below: On the verandah with my workmates, Esbank Terraces, Lithgow, 1983: (from left) Barry Downs, plumber, me, Matthew Retmock, apprentice carpenter, Grant Edwards, foreman, and Harry the local labourer.

Right: Kath Earl (left) with Judy on our wedding day, August 1984.

Below: With the Bilge Rat outside our job at the Shamrock Inn, Hartley Village, 1984.

Above: The day *Spirit* was launched, 1985: (from left) Pip, me and Sal. My two beautiful daughters are looking at another boat and another adventure.

Left: Boxing Day 1985, when *Spirit* was officially christened by Prime Minister Bob Hawke at the Cruising Yacht Club, Sydney. Bob, me and Benny Lexcen enjoy the moment.

Above: The Gang of Four, December 1985: Rod Muir, Tom Stephenson, Stevie Bakker and me (behind) prepare to go to Hobart on the *Windward Passage*.

Left: Leg II of The BOC Challenge from Cape Town to Sydney, 1986. It's bitterly cold in the South Indian Ocean.

Above: *Spirit* under the Triple M spinnaker coming up the coast of NSW, approaching the Leg II finish line, 1986. *(JEFF MELLEFONT)*

Left: Crashing away through the Southern Ocean, 1986.

Left: 'I didn't want to arrive like a wimp', but this is ridiculous – broaching *Spirit* as I came across the finish line.
(JEFF MELLEFONT)

Right: And at the press conference in Sydney, December 1986.
(SANDY HAMMOND)

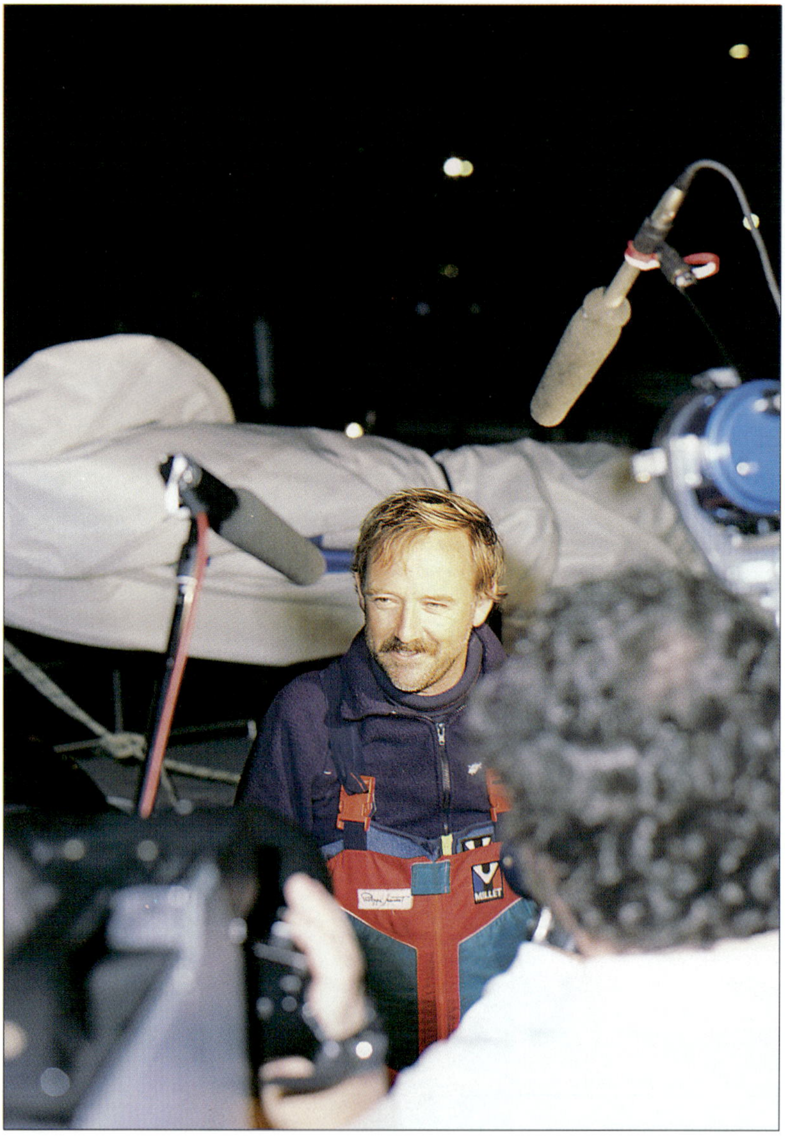

Above: On the dock in Newport at the race finish surrounded by media 156 days later!
(ONNE VAN DER WALL)

Left: With Judy and Max at home, 1988.

Below: Clean Up The Harbour Day, 1989.
(PETER BRENNAN)

Left: With Jack Earl the day the *Kathleen* arrived in Sydney Harbour after the restoration, 1991.

Below: Oh no! In January 1990 the house boat sank – at high tide the water came up to the roof line.

Above: At the press conference, 24 January 1990, to announce cities and towns that had taken part in the first Clean Up Australia Day: (from left) John Barrett, Sydney co-ordinator, me, Kim McKay and Giselle McHugh.
(PETER BRENNAN)

Right: Volunteers, Clean Up Australia Day, 1992.
(CATHERINE SECULA)

Above: The Clean Up The World team, September 1993. We had seventy-nine countries and 30 million people involved.
(CATHERINE SECULA)

Right: Aboriginal communities around Australia have always been involved with our activities.

Left: At the United Nations Headquarters, New York, September 1993: (from left) Kathy Giblin, our US representative, Jessica Fabrizio from the EGBAR Foundation, me, Cecile Alvarez, our co-ordinator in the Philippines, and Kim McKay.
(ALEX TOWLE)

Right: On the twenty-first anniversary of the Crash and Burn, 1994: Zoran Janjic and me (Zed is wearing my Cape Horn hat and I'm wearing my Crash and Burn blazer).

Above left: At Admiralty House on Australia Day just before being named Australian of the Year 1994: (from left) Jack Earl, Judy, Mum, Pip, me and Sally.

Above right: Prime Minister Paul Keating, Phillip Adams, and me at the Australian of the Year Awards.
(NATIONAL AUSTRALIA DAY COUNCIL)

Left: Tackling the assailant, David Kang, with Premier John Fahey on Australia Day, 1994.
(ROSS COFFEY/COURTESY OF WOMAN'S DAY)

Right: On the tour of honour as Australian of the Year, speaking to schools and community groups. During the year I gave more than 350 speeches.

Below: May 1994, the day we scattered Jack's ashes. On board *Maris*: (from left) Lani Tomaszewski, Mick Earl, Michael Tomaszewski, Tiare Tomaszewski, Maris Earl Tomaszewski and Ben Hawke.

Left: The crew of *Canon Maris*.

Below: Returning from Hobart, January 1995. Matt Tomaszewski and I enter Sydney Heads. *Maris* had never looked better.
(CATHERINE SECULA)

jutting slightly into the South Atlantic Ocean, but collecting the swells and currents from the Pacific Antarctic basin as well. The continental shelf rises steeply and irregularly, creating shallow areas that further increase the hazard to shipping. And, being a lee shore, the Horn is also frequently covered in mists and fogs. When conditions are bad — and they are never that good — Cape Horn veterans advise leaving the Horn well to port. But, despite the cold and the strong winds, the seas weren't particularly frightening for my passage, so I took *Spirit* inshore off Diego Ramirez Island and sailed quite close to the Cape, taking in this enormous and foreboding slab of land, framed by jagged snow-capped mountains.

I was surprised at how weak the Cape Horn light was, a pale yellow beam across the surging sea. Here I took out Jack's card, had a healthy slug of the cognac and consigned the clew-rings and thimbles from *Kathleen Gillett* to the deep. I raised the bottle again. 'To Kath, old mate.' I drank Jack's share too. I needn't have. On the very day I rounded Cape Horn, Jack had worked out my approximate position and posted me this letter for my arrival in Rio: 'As soon as we hear that you have sailed into Rio and completed the most hazardous section of your world-girdling voyage, I am going to get smashed and carry as many others with me as I can. We will overfill our jugs and drench our shirts and fill our boots and yell our congratulations across the bay.'

The morning after rounding the Horn, Bertie Reed came up on me from astern and passed within hailing distance. We were both jubilant at having made it, but we should have delayed the celebration. The worst was still to come. Between Staten Island and Tierra Del Fuego on the mainland there is a passage called Le Maire Straits where the current is fearsome. With Bertie Reed on my heels I wanted to pass through for tactical reasons, rather than sail to the east and set a new course to avoid the Falkland Islands. I approached it in the dark with fifty knots behind me and huge, square-topped waves breaking across the cockpit. With boat speed of thirteen knots and eleven knots of current pushing me back, I roared through in the night, getting dangerously close to the rocky coast of Staten Island.

At one point I almost cut and ran, but the tide changed, the current weakened slightly and I stuck to my course and squeaked through with the first light of day, leaving Bertie well behind.

Along the southern coast of Brazil conditions got lighter and lighter, making it an absolute bitch to get into Rio. There's a very strong aero beacon up on Cor Cavado near the big statue of Christ. You can see it clearly at night from more than 100 kilometres out at sea. I had this thing in my sights for three nights as I lolled around in near calm. Few things are more frustrating than to be stranded so close to the finish line, but I stuck to my task and a breeze eventually carried me into port.

In the early evening the entrance to the famous city was quite breathtaking, with the green lights of Copacabana's beachfront in the foreground and the big light up on the mountain dominating the skyline. It was like some sort of Catholic Disneyland! I crossed the line and was taken under tow. Race director Robin Knox-Johnston came aboard with a couple of cold beers, and Stevie and Susie were waiting at the dock with a few more. That night we discussed how far back in the field I'd slipped — sixth in the class, seventh overall. After the Sydney leg it was disappointing, but nothing like Cape Town. *Spirit* couldn't be written off yet.

Rio De Janeiro is an exciting place, but for me it seemed tainted because in almost everything you do there is the whiff of violence and the stench of corruption. Nothing gets done without a sling, and sometimes the queue for the handout stretches from heads of department down to petty officials. As we sailed around the world, some of us shared shipping containers between ports, in order to have spare parts, sails, tools, even toys like pushbikes delivered to us at our next landfall. The clearance and customs charges in Cape Town had been about the equivalent of $250 per container, in Sydney about $350. In Rio we were told by the officials that there would be some charges related to the containers. We said, yes, we expected that, how much would they be?

The man in the uniform looked very serious, pointed to one of the containers and gravely announced: 'This one will cost $US22,500.'

This was a joke, of course, but none of us was laughing. No one in his right mind would pay that kind of money to release a container whose contents were worth a fraction of the amount. The official knew this too, he was merely opening the negotiations. This was the government charge, but he knew someone who knew someone who, for a small consideration, might be able to help . . .When it got down to $2,000 a container we paid the backhanders and walked off in disgust.

Then there was the matter of personal customs formalities, for which we had to present ourselves at Customs House, wearing long trousers and collared shirts as a mark of respect, and wait around for endless hours in a hall full of people, some of whom had decided to make use of the time by catching up on some sleep. I spent several hours in there with Bertie Reed, who knew a bit about the Brazilian bureaucracy and wouldn't put up with a lot of it. I looked up from my magazine to see Bertie behind the counter, stamping our papers and babbling away in bad Portuguese while the officials went into a mild panic. It worked. We were out of there pretty quickly.

I decided I didn't need a long holiday in Rio and flew home for a couple of weeks, but when I got back to Rio I fell straight back into the arms of the corrupt bureaucracy. I had been awaiting the arrival of a new generator which didn't come in until five days before the restart. I had to get it cleared, but there was a problem, of course. It couldn't be cleared for six days. There was a British expat on the Rio race committee who'd been dispensing advice on how to deal with these situations, so I did what he'd suggested and started negotiations with an official, which meant working out his price. The problem was that other officials along the food chain somewhere sniff out that a deal is in progress and start to place obstacles in the way of it, angling for a split of the take. As the number sharing in the action grows, so does the asking price. In the end I got the generator in time, but I'd paid far more in bribes than the damn thing was worth.

The race was on again, this time with a difficult start in light conditions, resulting in a log jam and a bevy of protests. Titouan Lamazou came through with good boat speed, but I couldn't give him the buoy room he needed because I had another boat to weather. It was a near collision and Titouan fired off a protest against me and one other boat. I protested against the two boats that had caused my problems, and soon there were very few boats in the race not flying protest flags!

Despite this, we got into open ocean fairly well, and then came decision time. The Brazilian current flows south quite swiftly and to make good time going north you must either hug the coast or head out wide, leaving the current to port. This latter option is the conventional way, sailing into the southeast trades, then north to the equator. But the shorter route is to tack up the coast, using the sea breezes in the same way you would sailing up the east coast of Australia. This was a familiar style of sailing and the one I chose, sleeping as I tacked offshore to the current, then keeping watch from the cockpit as I tacked back towards the coast. Unfortunately, it wasn't a familiar coast.

I was down to a No 3 jib and a deep reefed mainsail on the onshore tack, going along quite nicely, when exhaustion got the better of me and I fell asleep at the chart table. I woke up to the sound of clacking, like Flamenco dancers stamping their heels. *Spirit* was hardly moving. I clambered onto the deck and saw lights all around me. Had I died and gone to a hell where they play the Gypsy Kings all day long over the PA? No, the clacking was the sound of fishermen starting their motors to head outside fishing with the coming dawn. I'd sailed right into a bay and was only minutes away from a beach. The fishermen seemed totally cool about this, obviously thinking I was coming into anchor, but in this peaceful setting I was suddenly panic-stricken!

I hurriedly checked the depth sounder, checked the chart for reefs and sailed away from the problem, expecting at any moment to hear the grinding of keel on coral. I was very lucky to escape onshore doze — the fate that has befallen several competitors in the BOC Challenge. In 1982–3 *Gypsy Moth V* had been smashed to pieces on Gabo Island, and Richard McBride had let *City of Dunedin* drift onto a beach in the Falkland Islands, while in 1986 John Biddlecombe had put *ACI Crusader* on the reef in Tonga during his qualifier, and in 1994 Jean-Luc Van Den Heede was to plough *Vendée* into the beach at Kiama. I have been woken by disturbances in peaceful bays many times since that dawn in Brazil, and I have never had a harsh word to say to the fisherman starting his diesel motor.

The boats that had hugged the coast were now making good time quite close to each other. I had Guy Bernadin in *Biscuits Lu* in sight as we approached a tricky section of coast with a narrow passage between it and a coral island. This island also had a coral shoal stretching seaward a long way. On my chart the passage was about the width of a pencil. I got Guy on the radio and asked him what he thought about it. He said: 'I have a chart of the passage. It looks okay. Follow me through if you like.'

My charts covered a broader area and had no detail on the passage. I told Guy I didn't have the stomach for it. I would go around. I tacked offshore and the island began to fall away. Then I tacked again to the northeast and, in my eagerness to keep *Biscuits Lu* within reach, cut the corner too fine. My chart showed a field of vigia, coral hazards that rise up in columns off the reef at uncertain points. As the depth sounder started to flick madly up to twenty feet (six metres), and then fifteen feet (4.5 metres), I realised I was in the middle of it.

There were two dry rocks shown on my chart and they were dead ahead, so I tacked again and nervously kept a watch on the depth sounder. Flick, flick, flick . . . Soon the flicks were almost constant and I knew I was in trouble. *Spirit* drew just over three metres and these coral heads were four or five metres below the surface and getting closer. I made a split-second decision to cut and run. I knew how easily you could lose a boat on coral, so I did what I had said I'd never do in this race. I turned the boat around and sailed back the way I'd come. For another nerve-wracking hour I played Russian roulette in the vigia, then the flicks of the sounder became less frequent and slowly the bottom levelled out. I had watched a fishing boat power way out to sea before tracking north again, and now I followed the same course.

I'd had my heart in my mouth more often than I liked for several days now, and believing that bad luck travelled in threes, I breathlessly awaited the next signal for disaster. I didn't have to wait long. It was a beautiful, moonlit night and I noticed on the radar that something seemed to be tracking me. It would appear on the screen, fall back and then appear again. I knew it wasn't one of the other competitors because I had their positions plotted. On closer inspection I decided it couldn't be any yacht, because of the manoeuvres it was performing on the water. Pirates in a power boat? It came closer and closer on the radar and I began to get quite concerned. While keeping an outward calm, I searched the cabin for a weapon to protect myself in the event of a boarding. Then it came up on my stern — a water spout!

I was still chasing Guy Bernadin as I approached the town of Vittoria. Here I could smell the familiar smells of the land, the city, humanity and the coffee which is the major industry. As I moved further up the coast another smell overpowered all these and I realised I was sailing through a massive sewer discharge. The spray was flying and it got so bad I had to wrap a towel around my face to keep the foul-smelling water out of my nose and mouth. Eventually I cleared the polluted area, but the next morning I was horrified to find the hull coated in a black slime. I felt both sick and angry, and the anger stayed with me for days.

CHAPTER FIFTEEN

BIRTH OF THE CLEAN UP

North of Brazil in the Atlantic Ocean you enter the Horse Latitudes, so called because this was where the early mariners began to slow down ahead of the doldrums, and had to throw their horses overboard before they drank all the water. This area is also known as the Variables, because you are just as likely to encounter the doldrums as a decent breeze, and within the Variables is an expanse of ocean known as the Sargasso Sea.

The Sargasso Sea was a place I'd read about as a kid, a place made famous in the logs of ancient mariners and in the mythology of the sea. One of the myths was that this was the nesting place of the halcyon bird, a species rather like the kingfisher who, according to the ancients, charmed the sea to tranquillity and bred during the winter solstice in a nest of floating weed. I loved all this stuff, and for me the place had a magic unequalled by any ocean other than the Great Southern.

Although the doldrums had been disastrous for me earlier in the race, I was picking the weather pretty well now, with the help of Clouds Badham's computer and by keeping a weather eye on Bertie Reed, who was close by again and knew how to stick on an isobar line. Bertie and I reached the front of the fleet a couple of times through here, and despite the heat and the calm, I was enjoying myself immensely.

This was the second time I had sailed into the Sargasso Sea. The first time, with my full concentration on beating the doldrums and staying in the race, it didn't have much effect on me. This time I was racing for the finish line in Newport, but I made myself take notice and the sea had an effect so profound that it was to change the course of my life. I can't overstate the depth of my disgust when I sailed into this pristine place of smoky blue water, dotted by semi-submerged golden weed so vital to the marine food chain, this sea of magic and myth . . . and found it littered with rubbish.

First a rubber thong, then a toothpaste tube, a comb, a plastic bag. As *Spirit* cut through the almost glass-like sea, propelled by the lightest of breezes, the rubbish popped up on both sides of the bow. I didn't know where it had come from — although the recreational cruise industry must bear some responsibility in those waters — but it took me back to the Ala Wai Canal that Sunday in 1975 when we cleaned up so much filth and thought for the first time about what we, as sailors, were doing to our playground, our oceans. And it took me back even further, to the Sydney Harbour I knew as a boy, and the one I knew forty years later, with broken glass at the tide line and plastic bags covered with green slime on the surface. And I thought about the big bag of garbage I had in the quarter-berth down below, garbage that would have been floating in the sea too, were it not for Mark Schrader's involvement with the Center For Marine Conservation.

What I saw in the Sargasso Sea helped put into perspective a number of issues I'd been thinking about for a long time. Perhaps for the first time, as I climbed through those latitudes, it occurred to me that if I could keep a bag of garbage down below, surely a million recreational sailors around the world could do the same. Each one of us just had to come to the understanding that one person could make a difference.

I came back to the here-and-now with a sudden change of conditions. Into the North Atlantic now, with a steady breeze from the east and the Gulf Stream to haul us along at a bonus seven knots. We were in a tight group at the front of the fleet now — me, Bertie, Titouan, Philippe, Guy Bernadin and John Martin. The last thing you wanted to do was screw up that close to home, but we were very tired, and ready for the finish line. And unfortunately the shipping lanes were the busiest of the entire course.

I was up and down all night, every night, trimming and navigating, running from chart table to cockpit and back to the radar. One night I picked up the lights of a ship dead ahead. I got it on the radar at about thirty kilometres off. I called on the International Distress Channel, channel 16. 'Hello ship, this is yacht *Spirit of Sydney* at [I gave my position], please give your co-ordinates.'

All vessels are required to keep a listening watch on this channel, and I got a number of responses, but none was the ship I sought. Then the voice of another skipper came on. 'I think I know the vessel you've got there. Greek freighter. Keep trying.'

I did and the ship kept coming. I was close-hauled and didn't want to give an inch. I kept plotting it on the radar until it got so close I ran back on deck. There it was, coming straight towards me. I held my course and put a spotlight beam over its bow. Greek name. If it was close enough to read, it was too close period. I cracked sails and ran away as the freighter passed by me a matter of metres away. If I'd been asleep it would have cut me in two. I jibed away and ran down the side of the hull very close in, flashing my light in the wheelhouse. Eventually someone came onto the bridge. Then I passed under the stern and read the name, *Angelina Aquiluca.* I wouldn't forget it in a hurry.

'Hello yacht, this is *Angelina Aquiluca.*'

'Hello yourself, you stupid bastard!'

I tore into the operator with all the fury of one whose life might well have been vanquished in the most senseless of ways. I advised him of parts of his family history of which he may have been unaware, then I described to him biological impossibilities that would be performed upon his person, should the good ship *Angelina* ever venture into Sydney Harbour. And for good measure, I asked him where he'd bought his skipper's ticket. I was still shaking with rage when I got off the radio, but I felt a little better for letting it out.

Soon there were other skippers on the line. 'Hey, great stuff, Aussie!' *Spirit* and her velvet-tongued skipper had become the main feature on the North Atlantic shipping lane that night.

I felt I had a real shot at winning the leg, which would put me in good shape for a third overall. So, with Newport in mind, if not yet in sight, I threw everything I had at the effort. *Spirit* flew, but not fast enough, and by the time I got really close to Newport, I'd had word that *Tuna Marine* was already across the line. Then it became a scramble for the minor placings as half a dozen boats streaked for the line. After so many miles and so many months, it was an extraordinary finish. Guy Bernadin was in sight and in front of me, so was Philippe Jeantot, but I had Bertie and Jean Yves Terlain tucked away. Coming from behind, Titouan Lamazou managed to finish ahead of me too, but it was incredibly tight, with four boats finishing in less than two hours.

I was fifth home in the leg, sixth finisher overall. I hadn't won, but all finishers in the BOC Challenge are winners. I'd had a shocking leg, a middling leg and two bloody good ones. The sword of possible failure was no longer hanging over my head.

The fast finish took a lot of people by surprise, not least of them Judy, who was still in the air en route to Newport as I was taken under tow. Stevie Bakker was there, with an understated nod of approval, and he had with him Gary Payten, one of my Lord Howe Island mates. It was a hugely exciting time.

Just incredible! Half an hour apart after 8,000 kilometres. As we went out to meet Ian you could see more boats on the horizon. It was even more thrilling for us because Ian had led the pack a couple of times in the latter stages of the leg. There was a sense of tremendous jubilation. The competitors had got very close to each other, of course, and there was a big celebration on the dock.

Kim McKay

We got to the dock and the international media were everywhere. Kim whisked me off to a press conference at the Sheraton on Goat Island, unshaven and still in my oilskins. I barely had time to don my black Akubra. The mood was one of jubilation, but the media wanted to know about the Rio protests. The end result was that they were all dismissed, but at that moment we could not have cared less about technicalities. We had sailed solo around the world, that was all that mattered.

There is a great relief when you have met a challenge and succeeded, but there is also considerable sadness. The camaraderie of the fleet, built up over so many long and lonely nights when the voice on the radio, coming from eighty kilometres astern or ahead was your only comfort, is hard to let go. The adventure is over, but the adventurers are not necessarily the same. While we waited for the rest of the fleet to arrive so that the presentations could be held, Judy and I flew off to Clearwater, Florida, for a holiday with Courtney Ross. I needed that time to come down off the high, to relax and prepare myself for the resumption of normal life. But I spent most of the time in Courtney's workshop, making crazy little visual gags for the 'Not the BOC Awards', an unofficial roast that Kim McKay and one of the other girls were arranging.

Kathy Giblin had worked on global promotion for BOC and we'd become friends. We thought it would be fun to put on a show for the competitors. It was one of those wonderful, spontaneous nights that you can never repeat. Ian has a wicked sense of humour and this night really brought it out. Titouan Lamazou, one of the great characters of the race, had a habit of crashing a car in every port, and he'd also managed to hit an uncharted rock at Cape Horn. He became known as 'Crash'. Ian built him a toy boatshed with cars and bikes coming through the walls. He'd signwritten 'Lamazou Brothers Yacht Repairs — Auto and Marine' across the front. Titouan was delighted. There was such a strong bond between those guys.

Kim McKay

Before the last of the parties I handed *Spirit* over to Stevie Bakker, the best Nigel you could ever hope to have. Then Judy and I partied under a marquee at Goat Island until after four in the morning, got up at seven and caught the shuttle to Kennedy to connect for Paris and London. I remember stretching out in the plane, trying to shake off a hangover, and suddenly realising that I didn't have one. I was trim and lean and, at forty-six years of age, the fittest I'd been since my football days. Physically and mentally, I felt strong and ready for new challenges. And I knew I'd have them too, because apart from the fact that my business was in a bit of a shambles, the Sargasso Sea was weighing on my mind and pushing forward the germ of an idea.

Kim had done a marvellous job on my public relations for the duration of the BOC Challenge, and I returned to Sydney with a much higher profile than when I left. This was something of a mixed blessing, but it certainly helped me get a foot in the door with some important people. Harry M. Miller's agency had organised deals for me to contribute newspaper articles on *Spirit's* progress in the latter stages of the race (I'd dictate them to Judy over the phone and she'd type them up and fax them), and now, through Harry's Speakers Bureau, I found myself much in demand as a celebrity speaker. In addition to this, as an Australian who'd achieved something on the world stage and would perhaps be an inspiration to others, I began to receive invitations to become involved in public committees and events.

I suppose if I'm totally honest about this, I have to say that I enjoyed the attention, but I had a real fear of becoming a one-hit wonder, a 'celeb' dragged out of mothballs every so often to recall one great moment in time when the planets aligned and, for a change, things went right. My life didn't start with the BOC Challenge, nor was it over when I crossed the finish line. The BOC was one challenge; there were plenty more.

But before I tackled them I had to re-enact the BOC for the television cameras. My mate Alan Morris from the Mojo advertising agency had lined up a Toohey's beer commercial featuring *Spirit* and me. I needed the money and it was a good opportunity to give Triple M some more value for their sponsorship money. We took *Spirit* down the south coast to Gabo Island, hoping to shoot the commercial in rough seas like those I had encountered in the Southern Ocean. Unfortunately the weather was beautiful, but they shot what they could. One scene involved me climbing the mast, which is one of my least favourite things. I climbed the bloody thing and the chopper flew past and took the shot. Back at deck level I got the radio message: 'Lovely but not quite right. One more time.'

Back up the mast, chopper flies by, shot in the can. That night over a few beers in Eden, I asked the cameraman what had been wrong the first time. 'Nothing,' he said with a wicked grin. 'I just heard you love climbing the mast.'

The rough-weather footage for the commercial was finally shot back in Sydney. The weather was still too good and the seas not rough enough, but the camera created the illusion of a stormy sea by shooting from the rocks while I surfed *Spirit* in the waves just metres off Bondi's Ben Buckler, rather as I'd done as a kid.

A few months after my return to Sydney I explained to Kim McKay my concerns about Sydney Harbour and the thoughts that had come to me in the Sargasso Sea. I suppose I confided in Kim before anyone else because I knew she was a great organiser and had a real respect for the marine environment. She had promoted yacht races and surfing contests, why not a cleanup?

When Ian came back from the BOC race he was a bit lost. That's normal. Anyone who embarks on an adventure like that is bound to feel a bit let-down, a bit empty inside when it's all over. One day he walked into my office and started talking about the ocean environment and what he'd seen in the Sargasso Sea. He'd been sailing that weekend on the Harbour and I think what he'd seen there had made him realise there was some urgency if he was going to do anything about it. He told me about the cleanup in the Ala Wai Canal all those years before. 'Do you think that would work here?' he asked. I said, yeah, I thought it would. It was a brilliant idea and I knew people would relate to it. If Ian felt this way, I was sure thousands of other people would too. I swung around onto my old IBM Golfball typewriter and started writing a letter to Laurie Brereton, who was then the NSW Minister for Ports, while Ian dictated his thoughts.

Kim McKay

Around this time John Singleton bought a houseboat at Pearl Bay and asked me to restore it. He said: 'Mate, I've never been happier than when I lived on a houseboat. You're a sailor and a builder, see what you can do with this.'

'This' was in fact a twelve-tonne barge used for ferrying automobiles and carriages across the Spit until 1910. Then, in the Great Depression there had been a tent village of unemployed at Pearl Bay and people built makeshift homes on any platform that would float. The previous owner had kept the house and, over a number of years, replaced sections of the punt so that it was a very solid affair. Now, Singo wanted me to turn it into a floating palace!

I got hold of the Bilge Rat and we went to work on it, virtually dismantling the house but retaining all of its attractive features. It had, for example, lovely carved panels, stained glass and ferry boat windows, so we brought in architect Alex Genko and incorporated all of this into a simple new design with two bedrooms, inside and outside living areas over two levels. It was a beautiful little houseboat and Singo loved it, not that he had it for long.

In the year or so after I came back from the BOC Challenge people were forever throwing real estate deals in my face, and I was inclined to listen to a few of them. The market had changed a lot since the '70s, I was a bit older and wiser and, perhaps most importantly, I was well clear now of the morass of debt that had almost engulfed me. Once I channelled my energies back into business, Alfred Builders started to do well too, but I was mainly on the lookout for some astute real estate purchases.

So after about umpteen approaches along the bar at the Mosman Rowers' Club, a bloke we knew only as 'Pinky' collared me and I listened. 'Mate, I'm with Equiticorp. We're looking for investments, put a deal to us.' I didn't say no, and Pinky kept coming back with the same line. Eventually we looked at an old factory site in St Peters on Sydney's industrial southside. I could see there was some value in the existing structure and that we could make it work if we infilled the building behind. I did a costing and we put a joint venture together with Equiticorp. We built a factory which was purchased by Emery Chemicals, realising a good profit for both of us.

I did one more deal with Equiticorp, but I found them a funny mob, particularly their boss, a flinty-eyed New Zealander named Allan Hawkins, who was one of the first of the '80s entrepreneurs to end up behind bars. I went off him in a big way when he advised Pinky to shave off his handle-bar moustache because he 'looked like a poofter'. I backed off and saved myself a lot of anguish, but that wasn't the last I saw of Pinky.

I was invited to crew on *Ragamuffin* in the 1987 Sydney to Hobart and I didn't need to be asked twice. A few weeks before the race I found that Pinky was also on the crew as cook. He nudged me in the Rowers' one evening and said: 'Just like old times, eh?'

If you can boil rice and open a can while leaning at thirty degrees, you pretty much qualify as cook in a Sydney to Hobart race, but Pinky had very different ideas. He went to a fashionable Mosman eatery called Providores and ordered huge quantities of gourmet food, which must have caused Syd Fischer apoplexy, although he didn't say anything. Then on race day he showed up with his hair dyed pink.

It was a good, tough race and we were second across the line after battling *Condor* all the way, and Hughie Treharne and me having to sew up a mainsail blown in Bass Strait. As we stood around the dock savouring our race experiences, Pinky announced that he was off to a suite at the Sheraton 'at Mr Fischer's expense'. We shook our heads in amazement, since we were all roughing it, and here was the bloke who had nearly burnt the boat to the water line with a galley fire trotting off to his luxury suite!

There was a sad epilogue to this. On the morning after Clean Up Australia Day, just over two years later, a body was found floating in Mosman Bay. It was Pinky, who'd apparently left the Rowers' in particularly good spirits, fallen over and knocked himself unconscious. He was an odd fellow, but well-meaning, likeable and not without a sense of irony. I'm sure he would have appreciated the sad laugh we had at the Rowers' when we drank to his memory that evening. Pinky's timing was terrible, someone suggested. If he'd gone a day earlier his mates would have been there to clean him up and he could have added his weight to the rubbish count.

During the construction of the Equiticorp factory at St Peters we had a record run of bad weather which threatened to blow out the budget. I was at the site every day, making sure we used every available moment, and began to notice a little blue cattle dog around the site. In fact it was rather difficult not to notice him. He'd greet you by jumping up and covering you with muddy paw prints. He had a nice big head, obviously intelligent, and I fell for him. It had been some years since Alf had died, and I guess I'd come out the other side of that mindset all dog lovers know, the one where you think you couldn't stand to lose another mate like that.

I asked around and found that he was a street dog who'd been adopted by a Maori family down the road. The kids had lost interest in him and he'd pretty much gone back to the street. His name was Max and the word was if I made an offer I could have him. I went home and told Judy.

'No way. Look at the size of this house! We can't have a dog.'

The next day I went to see the Maoris with a couple of hundred bucks in my pocket. I asked them how much and they said I could have the dog for a couple of slabs of beer. Deal done. Later on, the lady of the house brought him down the road on the end of a rope. I phoned Judy and said: 'Look, I know you said no, but what if I bring him in to meet you at the office?'

So off Max and I went for Friday evening drinks at John Singleton's office. I carried him up the stairs and he sat under the piano and glowered at everyone. Judy loved him. He slept in the backyard for a night, then she bought him a basket and blanket and he's lived inside our little house ever since.

In 1987 the journalist and sailing historian Bruce Stannard solved a puzzle which had plagued Jack Earl for many years. After eighteen months of painstaking research, Bruce, a pretty amazing bloke, uncovered the whereabouts of *Kathleen Gillett*, the beautiful ketch Jack had commissioned in the 1930s and had sailed around the world in 1947 and '48.

After Jack sold her in the 1950s and later commissioned *Maris*, *Kathleen Gillett* disappeared into the anonymity of a dozen Pacific ports. She hauled trocus shell under a Torres Strait Islander crew, went to work for crocodile shooters in Bougainville and the Solomon Islands and shipped copra in New Britain. Then in the early 1960s she was bought by Rabaul businessman Reg Stephenson, who restored her as a cruising yacht. In 1967 Stephenson sailed her down to Sydney to take part in the Hobart race, and before the start Jack went aboard and was happy with what he saw. But Stephenson's stewardship of the boat ended in 1976 when she was thrown up on a reef in Guam during Cyclone Pamela. Now Stannard, a mate of Jack's and a councillor of the Australian Maritime Museum, had located her again.

The old gaff-rigged boat, arguably Australia's most famous cruising yacht, had been partially restored by a Norwegian-American family living in Guam, but the years of neglect and hard work had taken their toll, and *Kathleen* seemed certain to end her days at the bottom of Guam Harbour. Stannard, using his reporter's instincts, seized on the Norwegian connection as a means of bringing her home to feature in the new Maritime Museum at Sydney's Darling Harbour. The current owners were Norwegian, Norwegian-born designer Colin Archer drew the lines on which *Kathleen* was built, Charlie Larsen who built her was Norwegian, and there was a rich Norwegian boatbuilding tradition in Australia, not the least of it being the Halvorsen family, who eventually would restore her. Why not ask Norway to purchase the boat and present it to Australia as a Bicentennial gift?

Through the offices of Per Haugestad, the Norwegian ambassador to Australia, and Trygve Amundsen, the consul general in Sydney, Stannard made approaches to the Norwegian royal family and found them happy to oblige, provided the current owners were prepared to sell. He then contacted the owners, Vaughan and Jane Tyndzik, who had acquired *Kathleen* in 1981. A letter came back from Mrs Tyndzik:

> Your call came as quite a surprise to Vaughan, but not so much to me. Somehow I always knew we might receive a call such as yours . . . To you the *Kathleen* is your history, your past. To us, she is our present and our future . . . Yet I see your side . . . Vaughan speaks about finding another boat, yet the way the *Kathleen* feels as we sail . . . no other boat could come close in my eyes or in my heart . . . When I hold her tiller and feel her strength I am secure, and I don't know that I could ever feel that way with another boat . . . I will leave the final decision to Vaughan, for I know he will do as he feels he must, not just as the owner but as a man who loves his boat and yet might find in his heart that the best thing for her is to let her go.

Bruce eventually negotiated a purchase price which was way above *Kathleen*'s market value, but the Tyndziks continued to vacillate. *Kathleen Gillett* had a powerful hold on people. The Zim Israel Navigation Company had agreed to bring the boat back to Australia free of charge, in return for Prime Minister Bob Hawke's outspoken support for Israel, and Stannard had organised to freight a cradle to Guam for the old boat to sit in on the ship's deck for the voyage home. But as the date for the loading of *Kathleen* approached, Stannard and Ben Hawke, who was making a television documentary about the return of his grandfather's boat, began to voice concerns about whether the deal would go through. Benny asked me to fly to Guam to oversee the operation.

The flight was a freebie, and beggars can't be choosers, I suppose. Nevertheless, twenty-seven hours' flight time, detouring through Auckland, Pago Pago and Honolulu, did seem a little crazy, given that Guam is directly to Australia's north.

Vaughan Tyndzik met me at the Guam airport and offered me a bed in his study. He was very keen that we should go sailing on *Kathleen* the next day. I felt that my time would be better spent ensuring that the loading operation was in readiness, but bearing in mind the reluctance of the Tyndziks to sell, I thought I'd better be sociable.

We drove down to the foreshore in the morning with the two family dogs yapping in the back of the pickup, and Vaughan pulled out an old dinghy and we rowed out to *Kathleen* while the dogs swam behind. I'd never seen the boat but I recognised her straight away. Her lines were unmistakeable. But the closer we got the more appalled I became. It was clear to me that she was in perilous condition. She had been baked in the tropical sun for too many years, had washed across a reef and had sat rotting in the mangroves for several years, until Tyndzik began his restoration.

Vaughan Tyndzik's heart was in the right place, but he obviously knew nothing about boatbuilding and his repairs had been somewhat agricultural. There was concrete in the bilges and everywhere the signs of rough jobs. She was rigged out with perished old racing sails. We went down below and I began to fear for our safety at sea that day. *Kathleen* was of robust construction, but she was desperately in need of attention. I asked Vaughan if he had a compass. No, he didn't. Surely a hand-held one? No, not even that. Radio? He shook his head. Engine?

'Of course. We have a diesel engine.' He unclipped a rickety old companionway ladder and pulled it out, and there was a four-cylinder diesel engine with the head off and the pistons full of rusty salt water. I began to take a keen interest in what was on board. With considerable relief I noted that we had about twenty litres of fresh water. With the two dogs now on board, I figured we at least had survival rations. There was a huge life raft cannister on the fore deck, but I later discovered that there was no life raft in it. The upturned cannister lid was as close as you got to a raft.

Despite all this, *Kathleen* sailed beautifully and I could see why the Tyndziks were reluctant to part with her. However, I was able to convince Vaughan that letting her go was the only way he and Jane would survive. Two days later we sailed her up to the head of the bay, lifted the rig out and towed her back up to Guam Harbour, which was a fascinating graveyard for old, tired boats. I sawed the rotting bowsprit off and began the preparations for loading. There was more to this than simply supervising a team of workers. There was virtually no budget for this operation, so I sought a meeting with the Governor, Joe Ada, and the port authority of Guam, to request their co-operation and support. I laid it on pretty thick, telling them that we shared a common bond as island people, it was just that our island was a little bigger than theirs. I must say the people of Guam were fantastic, even though their port only played a bit part in the saga of *Kathleen Gillett*, and I made sure that we got a Guamanian flag to fly on her.

Finally we lifted the boat into the steel cradle, blocked it up and dogged it down, and she swung on board the Zim Line freighter. There were some tears on the dock when she left, but I think in the end the Tyndziks felt quite proud of their involvement. I am sure they would feel doubly proud now, to see *Kathleen Gillett*, beautifully restored by the Halvorsens and floating in a place of honour outside the Maritime Museum.

Some time later, Bruce Stannard again sought my assistance on behalf of the Maritime Museum. The museum was nearing completion by this stage, housed in an impressive Phillip Cox-designed building. Stannard said: 'I'd love to finish it off with a boardroom that has a real maritime feel to it. Can you help?'

It was the sort of project I could never walk away from, a challenge involving the two great passions of my life, building and sailing. Stannard had lined up funding from the NSW Timber Advisory Council to produce a showpiece using plantation eucalypts, usually thought of as inferior wood used for telegraph poles or cattle yards. I knew the quality of the wood and had no trouble with using it. Bruce had also lined up a design from interior architect John Davidson.

So, with the funding, the timber and the design in place, I flew to London and spent some time examining the finishes used on the *Cutty Sark* in dry dock at Greenwich. Back in Sydney I set up my joinery shop in Pyrmont and we set about a complete fit-out which included wood panelling, flooring, ceiling, table, chairs, galley and detail. I was particularly proud of some of the features, including the acid-etched glass skylight incorporating a Jack Earl pattern (Jack also did two paintings for the walls) and the swivel chairs modelled on a chair which washed ashore at Fremantle off a square rigger which had been wrecked on Rottnest Island. It is a room in which you can smell the salt air, and it does my spirit good any time I go there.

Since the Riana Two to Port race in 1985, the Short-handed Sailing Association had gone from strength to strength, mainly due to the boundless energy of Don McIntyre. It was Don again who in 1988 suggested we celebrate the Bicentennial with a two-handed race around Australia. We found a sponsor in Goodman-Fielder and set about the time-consuming work of organising the race, which was to be conducted in stages, with short intervals between each in which competitors could rest and repair their boats if necessary. In effect, it was a mini-BOC Challenge, but coastal racing can be fraught with dangers, so we paid particular attention to the safety aspects. As a member of the race committee, I worked with Don in formulating a race plan that covered every contingency. Or so we thought.

The race attracted a good fleet. Kathy Hawkins and Ian Johnston entered their trimaran *Verbatim*, Whitbread veteran Peter Blake came across from New Zealand with his multi-hull, David Adams was in it, and we had many fine cruising boats. The race began at the Heads with the first stage to Mooloolabah, and the first night out a ferocious gale blew up. Peter Blake said later that conditions were among the worst he had ever encountered in the Tasman, and that he had almost flipped his trimaran over.

The first disaster occurred when a boat started sinking off Port Stephens. The Water Police launch went out to rescue the two sailors, but it sank too under controversial circumstances. (When the police commissioner attacked the race committee over its emergency planning, it was revealed that the Water Police had apparently been a little remiss in this area themselves, leaving a hatch open which took water.)

Then came the real tragedy. Another boat hit Mermaid Reef off Forster and a competitor was lost. These things happen in ocean-yacht racing and the weather is a factor you simply cannot account for, but there was no doubt we were in the hot seat, and Don McIntyre was under intense media pressure. The race continued despite the tragedy, with Peter Blake eventually winning, and at a coronial inquiry into the death, Don and the Shorthanded Sailing Association were both exonerated. But it was not a high point for us, and there have been no more races around Australia.

We had finally got a response from Laurie Brereton over the cleanup proposal, but it wasn't the one we wanted. Our letter had simply suggested the creation of a citizens' committee to work with the government to establish a cleanup day, like the one we'd had at Ala Wai. Brereton replied that cleaning up Sydney Harbour was an excellent idea, and that his department had made a firm commitment to the people of Sydney to do just that. In fact, no one had cleaned up the Harbour since 1902, when the Colonial Secretary reported to Parliament that between 4 January and 30 June the Harbour Trust had 'removed from the Harbour and destroyed 1957 rats, 784 cats, 968 dogs, 225 bags of meat, two bags of fish, 1150 fowls, 25 parrots, 19 sheep, 11 pigs, 1 bullock, 5 calves, 3 flying foxes, 2 goats, 2 sharks, 118 rabbits and 10 bags of chaff'.

Brereton's letter some eighty-seven years later was a pat on the head, combined with a none-too-subtle attempt to make political mileage out of an issue that was, by definition, bipartisan. Pollution doesn't care which party you vote for. It really got me going because this sort of complacency was, to my mind, the cause of the problem. For too long we, as individuals, had sat on our bums and expected someone else — the ubiquitous 'they' who, you hear in pubs all over the land, 'oughta' be doing something — to fix our problems. I'd never been a political radical, but I'd always believed that governments would listen to the will of the people, eventually. On an environmental issue like this, however, it seemed to me that rather than taking to the streets with placards, we needed to be taking to the beaches with buckets and garbage bags.

I never have liked noes very much, so we started to look for ways around this minor setback. During the BOC Challenge, *Spirit of Sydney* had flown the flag of the Mosman Rowing Club, and partly because of this, Mosman Council had endorsed me as Mosman Bicentennial Citizen of the Year. I suppose, given the gongs I was going to receive later in life, this was nothing to get terribly excited about, but it was certainly better than the various orders of the sword that the finance companies had bestowed on me fifteen years earlier! And it opened a few doors in Mosman Council.

I went to see the new mayor, Barry O'Keefe, and told him that with his endorsement I would organise a day in which the people would clean up Mosman's seventeen harbourside beaches. He thought the idea had merit, but he doubted that I could stir the citizenry into such a selfless act. Nevertheless, O'Keefe wrote a mayoral minute which won the unanimous support of Mosman Council. So now we were official.

We quickly put a committee together, consisting of Kim and me and advertising mates John Singleton and Alan 'Mo' Morris from Mojo. Alderman Rod Jones from Mosman Council also came on board, with John Henderson from Mojo, who has stuck with us all the way. Much of the credit for the entire Clean Up movement must go to the members of that original committee, who recognised that cleaning up the harbour beaches was relatively easy if you could convince the people to help. Convincing people to participate was what the challenge was all about, so I used the 'old mates act' to secure a committee of communicators, including a couple of the best in the country. Singo agreed to write the ads and get them on air for nothing, Mo would write the jingle, Mojo would provide us with an office and Kim and I would start canvassing for volunteers.

After one of our early meetings, Mo phoned me and said: 'Bugger Mosman! Why don't we clean up the whole Harbour?' This had been my original and rather naive intention, but without government help it looked like being a logistical nightmare, so I had quite happily down-scaled to Mosman. I thought about it over a weekend sailing *Maris* around the Harbour, and on Monday morning I phoned Mo back.

I said: 'Yeah, you're right. If we're going to do it, let's do the bloody lot!' Clean Up Sydney Harbour Day was slotted for 8 January 1989 and we set about the business of letting everyone in Sydney know all about it.

But we still needed money to make it happen. I confronted the committee with this problem and Singo said: 'Mate, you know Peter Ritchie from McDonalds?' I said I didn't. 'Lemme get him on the phone,' Singo said.

The conversation, from Singo's end, went something like this: 'Pete, mate. Mate, Pete . . . You know Bicky? Bicky bloody Kiernan. Mate, bloody beauty, round-the-world yachtie, mate . . . Yeah, mate, that one . . . Cleaning up Sydney Harbour, mate . . . needs a bit of bloody help . . . I'll put the bastard on . . .'

Peter Ritchie was a former rugby player, and if there's a religion stronger than sailing, it's rugby. He went back to his people and came up with $25,000. We had three months to go, we had a major sponsor, we had a dedicated team and we had people screaming down the phone that they wanted to help. It looked like we had a cleanup!

The further we got into preparations for Clean Up Sydney Harbour Day, the more exciting it became. A group called ORRCA contacted us and wanted to be involved. They organised the rescue of whales and dolphins. Right across the community it seemed there were small groups of people who, for very different reasons, thought this was the right idea at the right time.

Having said we'd clean up the entire Harbour, we had to investigate just how big that task was. I enlisted the help of Kathy Muir to get us *Windward Passage*'s huge chase boat, which did about thirty-five knots, so that we could survey the area to be cleaned. To our considerable alarm, we discovered that there were 270 kilometres of foreshore in Sydney Harbour and that almost every cove and bay had a pollution problem of one kind or another.

Mo wrote a jingle which immortalised the words 'Yukky Yukky Poo'. It was an advertising campaign which had a direct impact on kids, simplifying for them the vague concerns they had for the future of their environment. They didn't know much about greenhouse gases and the hole in the ozone layer, but they knew what yukky poo was, and the man on the telly was saying the Harbour was full of it. So come on, Mum, come on Dad, let's go and do something about it! These guys knew how to win the hearts and minds of the community, and the campaign was an unqualified success. At the Mojo offices there were thirty-two incoming lines and this wasn't enough to handle the offers of help. And these guys were trying to run a business as well!

> *The support we got from the media was phenomenal. Triple M was hugely supportive, of course, but the help of the* Sydney Morning Herald *really got our message across. Throughout the early part of the summer, with hard news winding down for the Christmas break,* Herald *editor John Alexander ran regular articles with a special 'Cleaning Up The Harbour' logo. Then, on 7 January, the day before the cleanup, Paul Bailey and Alan Tate of the* Herald *broke the story about alarming toxicity levels in fish caught off Sydney near the ocean outfalls. This really brought it home to people. The time had come to do something about water pollution.*
>
> Kim McKay

CHAPTER SIXTEEN

CLEAN UP THE WORLD

As Clean Up Day, 8 January 1989 approached I began to get a nervous knot in my stomach. Did we really know what we were doing? Maybe Laurie Brereton had been right, we should leave it to the experts. The task seemed so enormous, and I had two fears: one was that no one would volunteer and we'd look like bloody idiots, having made such a huge fuss about it; the other was that too many people would turn up, there wouldn't be enough bags to go round, kids would wander off along dangerous foreshores . . . Think of a black thought and I had it.

I wasn't really confident that 'Yukky Poo' had done the trick until we held a briefing for area organisers at Mosman Town Hall a few nights beforehand. The place was packed and the air crackled with excitement. The fever pitch enthusiasm was somewhere between a revivalist meeting and an Amway convention, but here we had a hall full of people who were psyched up because they felt empowered by nothing more than their own actions. They were getting off their butts and doing something about a community problem, and by Christ it felt good!

I suppose that kind of energy could easily be misdirected (one thinks of a lynch mob stirring its own righteous indignation to the point where ordinary people condone an act of horrific violence), but in this case it was totally positive. There was a great joy about the preparations. The attitude seemed to be, this is our Harbour, our responsibility.

And so it was on the big day. We set up a 'tally room' in the Mosman Town Hall and from very early in the morning we were getting reports from our organisers of huge crowds turning up. Some people were working for a couple of hours before breakfast and other engagements, while many families were arriving to make a day of it, with a big picnic lunch to follow when the work had been done. I did a quick motor tour around some of the sites and I could see that this was going to be bigger than any of us had anticipated.

For the television and newspaper cameras we'd organised a boat tour which I labelled the 'Cuckoo's Nest Tour'. And it was quite bizarre to see so many people joyfully trooping through the water, often up to their armpits in disgusting rubbish, and loving it! There had never been a community event like this in Sydney's history, perhaps never in the world, and no one was more surprised than I was.

While it was fantastic to have the support of so many volunteers, the strain the sheer numbers created on our backup was enormous. We had the support of the Maritime Services Board and the Royal Volunteer Coastal Patrol and found that we had to constantly call on both as the day progressed. Right around the foreshore we had barges stacked with dumpster bins so that the rubbish could be removed without having to drag it through beach and bushland, but some of them became so overladen they looked like they might sink. And then one started to! The MSB responded to an emergency call and righted it before we lost a single bag, but the incident did raise a serious issue we had to confront. No matter how much rubbish people collected, we had to get rid of it before the next morning or we would have a public relations disaster on our hands. We had free access to all the council tips, but our fleet of trucks and barges was quite inadequate.

Meanwhile, we had a room full of city and national media

wanting to know how much rubbish we had collected. The previous night I had been asked by several reporters to estimate the haul. I had thought about this a lot and wondered whether I should be conservative. Nah, bugger it! I bravely predicted we would collect 100 to 200 tonnes of rubbish. Now, in the late afternoon at Mosman Town Hall, Kim McKay handed me the report. I blinked a couple of times, gulped and read out the statistics:

'The people of Sydney, or 40,000 of them, have collected more than 5,000 tonnes of rubbish, including more than 3,000 syringes . . .' When the media applaud you know you're on a winner, but applaud they did. '. . . and there are still many teams of volunteers at work and we expect to improve on that figure.'

There was no time to stand around congratulating each other. A report came in that there were hundreds of bags under the bridge at the Spit and the tide was about to wash them away. I had my Alfred Builders trucks out doing the rounds, so I got a couple of them to meet me there. We worked in waist-deep water to get the bags onto the trucks. Then, just on dark I got back to the tally room to find we had another situation in Drummoyne, where the streets were full of our rubbish bags. We had forgotten about pick-ups in Drummoyne, and the concern was that if the bags were left overnight people would throw their household rubbish on top and we would have twice the job.

I was scratching my head about this, trying to contact one of my own trucks, when a bloke called Harry phoned the tally room. 'How youse goin' with that Clean Up, mate?' I told him we were going well, but that we had a few problems. Harry said, 'No worries. I got an eight-wheel bogey tipper, I live at Silverwater and I'm on me way.'

Harry was an unlikely angel of mercy, but he stuck to his task until well into the night, driving his tipper up and down the streets of Drummoyne while we ran behind, throwing in garbage bags.

It was getting on for midnight when I finally ripped the top off a cold beer. I reckon I'd lost a couple of kilos since dawn, and I certainly felt a lot lighter. I said to Kim and the others: 'We must never do this again.' But even then we knew we would. It was too late to turn back now.

On the Tuesday after the Clean Up I took some of the main organisers out for a day on the Harbour on *Maris.* It was a beautiful, sparkling summer's day made all the more so by the fact that we could point to strings of pristine harbour beaches and know that they hadn't been that way a few days earlier. In fact, there'd already been a rainstorm since the Clean Up and the drains had begun to empty the next generation of crap back into the harbour. But that was okay, that was positive too. To us it underlined what we'd already begun to suspect — that you can't clean up the place once and leave it at that. We'd motivated 40,000 people to get off their butts and make an effort, but it wasn't really about picking up some rubbish. It was about taking responsibility. It was about giving a stuff.

The Clean Waters Act some years before had gone some way towards ridding the Harbour of industrial waste, but what we'd pointed to was the people's pollution, the crap for which there was no one to blame but ourselves. And more than half a decade down the track, I regard 8 January 1989 as the day that Sydney Harbour turned the corner.

> *We'd put together a very successful marketing campaign, not marketing a product but marketing an issue. Ian was the perfect person to front this campaign because he had no axe to grind, he was believable and he was a champion. People listened. But of course the timing was perfect. There was already a huge amount of concern about the sewage pollution on Sydney's beaches. Dr Sharon Bader, in her book* Toxic Fish and Sewer Surfing, *partially attributes the triggering of community awareness of these problems to that first* Clean Up Day.
>
> Kim McKay

The publicity in the wake of the Clean Up was huge too. The phones at Mojo started to ring off the wall again, and we had requests from communities all over Australia wanting to know how to organise a cleanup day. The Day of the Derwent in Hobart was the first, then the people of Wollongong and Port Kembla organised Project Armada, which brought Lake Illawarra back to life by pulling from its depths an astounding 158 cars and two buses! The mayor of Darwin, Alec Fong Lim, was on the phone next, and we helped him get Clean Up Fanny Bay started.

Without really trying, we had gone full-time into the cleanup consultancy business, which was fine with me, because if we'd stumbled onto a successful formula, then I was only too happy to have people use it. But it meant we had to look rather seriously at the structure of the Clean Up organisation, and to its level of sponsorship.

The Clean Up committee became a board, and we decided that we would organise a national cleanup day for January 1990, if we could co-ordinate a minimum of twenty community cleanups around the country, and find the sponsors to pay for it.

At this point John Singleton told me there were too many meetings in his life. He said: 'I'll still do the ads, but I'm on the sidelines unless you really need me.' That was fair enough. Singo has done more than most to make Australia a better place, and I'll always be grateful for his commitment and expertise, which enabled us to take the cleanup message to the people in the first place.

Kim McKay had met an extraordinary young woman named Giselle McHugh, daughter of Federal Labor politician Jeanette McHugh, now Minister for Consumer Affairs, and she suggested that Giselle join the board of our fledgling organisation. Giselle had a strong commitment to the environment, she was a good negotiator and she had honed her management skills during a stint with the rock band INXS at the height of their international fame. She proved to be an asset almost immediately, and together, Kim, Giselle and I got out on the sponsorship trail.

I was not always good at this. On one occasion I handed the man from Apple Computers a proposal meant for IBM.

Fortunately he had a sense of humour. But we wore away at the corporate world and we found big companies who liked the concept and liked our enthusiasm.

The first to respond positively was Westpac. They were looking for a national activity which could actively involve their staff, and they found the environmental and community participation benefits of Clean Up very appealing. The sponsorship and public relations manager, Alan Tippett and his lieutenant, Janine Bavin, were tough cookies to deal with, but they saw the challenges and benefits to the community the same way we did.

So Westpac became our major sponsor and others soon followed. Westpac committed $200,000, while Telecom, Ansett and Preen came in with varying amounts between $50,000 and $100,000 in cash or kind. We even had sponsorship from Harris Coffee and Tea at one stage, when they pledged one cent from the sale of every unbleached tea bag sold up to $35,000 in what was probably one of the first examples of cause-related marketing tried in Australia. By September 1989 I was able to go on television and announce that we had sponsors and commitments from twenty-three different communities for the first Clean Up Australia Day. We set up in Kim's public relations office in Mosman and went to work. Before the end of the year the Clean Up had grown to include 211 cities and towns around Australia, and our little group had become an important and influential community-based environmental group.

With Clean Up progressing under the stewardship of Kim and Giselle, I turned my attention to the next BOC Challenge in 1990–1. Two years earlier I'd sworn I'd had enough. Sailing *Spirit* around the world had been one of the highlights of my life, but it had cost me around $200,000 which I was still paying off, it had been tough on Judy and, although I found it immensely exhilarating, it had been tough on me too. If I went again, I'd be turning fifty as I ploughed through the Atlantic Ocean between Newport and Cape Town. Was that what I wanted at this stage of my life?

The answer, I'm afraid, was 'yes'. I'd finished sixth in 1987 in the company of some of the best solo yachtsmen in the world. With the experience I'd gained that time, and with a better boat, I reckoned I could win it. *Spirit* had, in fact, turned out to be a difficult boat, almost impossible in the light equatorial conditions. People who had sailed her subsequently had expressed amazement that I did as well as sixth!

> *The first real sail I had in* Spirit *was after the BOC when Ian chartered it and we entered the Australian Two-handed Series. We were leading the race up along Sydney's northern beaches, then we had a bit of light air up around Lion Island. We sat in the lee for four bloody hours while the entire fleet sailed past us! After that my respect for Ian increased enormously. How he actually got that thing around the world, I'll never know. On sheer determination and strength of character, I'd say.*
>
> Tommy Stephenson

There was another compelling reason for me to consider doing the BOC Challenge again. Over a few short months I had become known as one of Australia's leading environmental crusaders. People knew my face from hundreds of television commercials and they knew what I was on about. Even kids in nappies knew I was 'Captain Yukky Poo', the bloke who cleaned up Sydney Harbour. The speaking engagements were pouring in, people all over the country, all over the world, wanted to be a part of the movement we had created. Why not use this momentum to launch an environmental challenger?

I had built up a small property portfolio again — mostly inner-city commercial buildings — which I put on the market to raise money, and on 9 May 1989 I announced that I would be a challenger in 1990 and that I was seeking $1.5 million in sponsorship to build a new boat and establish a children's foundation to educate them about the ocean environment. My idea was that the new boat would be the flagship for this foundation, with school children from around Australia charting my progress and monitoring pollution levels in all the seas I crossed. The Worldwide Fund for Nature had already announced a similar scheme called Oceanwatch involving 100,000 kids and another 50,000 American kids were going to focus on the race as a school project, so my idea dovetailed neatly into this.

The more I discussed this plan, the more enthusiastic I got about it, but unfortunately the potential sponsors didn't seem to share my enthusiasm. One of the problems was that five BOC entrants from Australia were knocking on executive suite doors asking for money, and the others had pretty compelling arguments too! My mate Don McIntyre had called his boat 'Sponsor Wanted'. By the time the race started he had changed it to 'Sponsor Wanted Urgently'.

But perhaps the biggest problem was the recession. The heady days of the 1980s were over. An even bigger downturn in the economy than the collapse of the building industry in 1974 had gripped the nation. That year Bond and Skase both fell from grace. Rod Muir had long since chucked in the corporate life to grow roses. Every pinstripe suit in Sydney, Melbourne and Perth was ducking for cover. Sponsor a yacht? Turn it up, mate. What do you think it is? 1985?

Clean Up Australia Day 1990 was a phenomenal success, involving more than 300,000 people. During the year leading up to it I had been pressured to align the Clean Up with various political factions, which was totally against our philosophy. I responded that the only way a politician was going to get involved with this campaign was 'if he's down at the water's edge with gumboots and a rake'. When Prime Minister Bob Hawke officially launched Clean Up Australia Day, I presented him with a rake and a pair of gumboots, which brought the house down. (Hawke was actually extremely supportive, as was NSW Premier Nick Greiner.)

We moved the tally room to Darling Harbour's Sydney Convention and Exhibition Centre and set it up like the national tally room in a Federal election. It was tremendously exciting to see the numbers grow, often in the most unlikely places, like Christmas Island in the Indian Ocean. Perth had assured us that the Cleanup wouldn't work because their city was already spotless. Well, it wasn't. In this, the first national cleanup day in any country, ever, volunteers collected more than 15,000 tonnes of rubbish.

In January 1990, eight months from the start of the BOC Challenge, I had to square up to the fact that I wasn't going to get sponsorship. Eight years earlier I'd been prepared to take *Maris*, but not now. If I was going to sail, I wanted to win. I quietly announced that I was withdrawing.

On Australia Day I was invited to the awards presentation lunch at Government House, where Bob Hawke announced that Test cricket captain Alan Border was Australian of the Year for 1989. As I left the lunch I was cornered by a couple of television news crews, who thrust microphones in my general direction and started peppering me with questions. I thought the attention a little strange, since I was merely a spectator at the awards.

'Mr Kiernan, are you disappointed?'

'No, why I should I be?'

'Well, for it to have sunk like that?'

Sunk? What were they talking about. Suddenly I twigged. They meant my BOC entry, that was it. Yeah, I suppose my boat had sunk, in a manner of speaking. I got into the swing of the interview now.

'No, not disappointed at all. These things happen and quite frankly, it's no big deal. I'll just move on to the next challenge and never give it another thought.'

'But what about the danger?'

'Oh, I don't give a damn about that,' I laughed, still without thinking.

I prattled on a bit more, then the reporter said: 'You haven't heard, have you? The houseboat you built for John Singleton sank last night. People had to jump for their lives.'

Singo had long since disposed of the houseboat, and apparently the new owners had made some rather radical modifications which involved removing a hull plug, then overloaded it for an Australia Day party. Their insurance company got in touch with me after it had been raised by marine surveyors, and the Bilge Rat and I went to work on it again.

At Pyrmont, in Sydney's inner west, we had had an excellent response to Clean Up Sydney Harbour Day and an even better one the following year when it became Clean Up Australia. One day, not long after we had moved our Clean Up Australia office into an old stone terrace in Pyrmont, some members of the resident committee, led by a cheerful, robust woman named Jean Stewart, came to see me with a problem. The 'City West' redevelopment program was rapidly transforming their working-class suburb and in the process the last park their kids could play in, the James Watkinson Reserve, had been bulldozed. Could I help?

I got onto the Department of Planning and was informed that there were plans to create four or five new parks in the area, but these were some way off. In the meantime, the kids would just have to play in the street. I didn't think this was good enough and nor did the committee. I said: 'Bugger it, let's build a park ourselves!'

We found a lantana-covered block owned by the Maritime Services Board. No one could remember it ever being used for anything, so we moved in. I got some mates to donate topsoil, trees, bricks for the fence, tools, equipment, everything we needed, and over two weekends we built a fantastic park. We called it Interim Park, because it would have to do until the government built another one. Then all hell broke loose. The bureaucrats told us we were a bunch of 'urban guerillas', but we held firm and eventually we were granted a peppercorn rent. The James Watkinson Reserve is now being rebuilt and it's going to be a beautiful park, but they're not getting Interim Park back until the kids of Pyrmont get all five parks they've been promised.

In January 1991 I was back at Government House for the Australia Day Awards and this time I was delighted to accept the

Event of the Year Award for Clean Up Australia Day in 1990, sharing the honour with the 75th Anniversary Reunion at Gallipoli. One of the old Diggers came up to me at the luncheon to congratulate me. He said: 'If you show Australians a bit of leadership, they'll follow you anywhere.' I reckon he had the secret of the Clean Up's success in a nutshell. Whereas other public awareness campaigns tried to tell people what to do, we'd simply given them an idea and they'd done all the hard work themselves.

The second national Clean Up Day that year attracted 320,000 volunteers in 450 cities and towns, who collected more than 20,000 tonnes of rubbish, including 2,000 car bodies. We also introduced the first Rubbish Report, a statistical analysis of the rubbish collected which proved to be invaluable in examining our national litter habits.

Clean Up Australia was starting to become nationally owned, the biggest community participation event in Australia, and that year I was awarded the Order of Australia Medal, given an Advance Australia Award and named 'Communicator of the Year' by the Public Relations Institute of Australia. Naturally, I was thrilled to be recognised in this way, but I knew I was accepting the accolades on behalf of a dedicated group of volunteers who worked selflessly all year round to make sure the Clean Up ran smoothly. If the idea of Clean Up Australia had ever been the sole property of Ian Kiernan, that day had long since passed.

Westpac had continued its major sponsorship of Clean Up Australia and during 1991 I had a call from someone in the bank's head office. 'Ian,' he said, 'we have a guy from the United Nations Environment Program here and he'd very much like to meet you.'

Kim McKay and I had a meeting with the UN representative and explained to him the simple principle by which Clean Up Australia worked. After a while he cut in and said: 'Hey, you guys are amazing. You actually do things!'

As a result of this meeting, Kim and I were invited to visit UNEP headquarters in Nairobi, Kenya, in June 1991 for discussions with Dr Mostafa Tolba, the director-general of UNEP and his staff. The purpose of these discussions was to work out a means of applying the principles of Clean Up Australia to a global initiative.

Kim had written a proposal on how she thought Clean Up the World could work, and as we walked into Tolba's imposing office we knew we had to use every skill we had developed to deliver the message professionally, particularly when we looked across the desk and saw the director-general's red pen comments scribbled all over our document. We felt a bit like we were back in school.

The basis of the proposal was that we would produce 'how to' information — books, videos, posters and other materials — in a number of languages and make them freely available to bona fide community, government or environmental groups around the world. It would be up to the local communities themselves to decide if they wanted to participate, and we would help them promote their activities. We would share our Clean Up formula with people all over the world for free, in a spirit of global co-operation — more or less like franchising the Clean Up without being paid for it.

Out of this historic meeting Clean Up the World was born, with 17–19 September 1992 (the best time of year climatically) set as the date for Clean Up Day all over the world.

UNEP committed $US100,000 in seed funding to enable us to spend time over the next year developing the campaign and to search for global corporate sponsorship. It took months of negotiations with the UNEP bureaucracy after that time to finally sign a Memorandum of Understanding to get the money they had pledged. Then, at the last minute, just as we were about to sign, they imposed a new condition on our agreement. They would only give us the money if we could guarantee a matching seed-funding contribution.

Getting that kind of money from a local organisation to seed-fund an idea in the depths of a recession seemed impossible, so Kim grabbed the bull by the horns, as is her style, and immediately flew to the United States at her own expense. There she teamed up again with her former BOC Challenge colleague Kathy

Giblin, and they found the money within a week from the California-based EGBAR Foundation. (EGBAR stands for 'everything's gonna be all right', a very Californian attitude.) We had our matching funding and we were in the global cleanup business.

Giselle came back on board to help run the campaign and we spent about a year assessing worldwide interest in the project. We were overwhelmed. Communities everywhere wanted to be involved. It didn't matter if they were from a developed or developing country — people everywhere just wanted to grapple with the problems themselves.

But the international business community wasn't as enthusiastic about our idea as the people were, and we had to postpone the inaugural event until September 1993. By now Kim and I had formed a strong partnership in Clean Up and had perfected a pretty good banter in our presentations. We occasionally came on somewhat evangelical, and we joked that we were becoming the Jim and Tammy Bakker of the environment.

We knocked on literally hundreds of corporate doors in the US and Europe, but most meetings just ended with the 'suits' looking bemused and asking how on earth we thought we could organise a global activity from 'down there in Australia'. Assurances that fax machines and computer networks had overcome the tyranny of distance fell on deaf ears internationally, and it wasn't until American Express in Australia came in with another $100,000 that the campaign was assured.

Flying home from one of our CUW meetings in Europe in 1992, I reflected on how far we had come in just three years, and how much further we would go in the next three. The Clean Up had become the driving force in my life, and it was time to acknowledge that and do something about it.

Up to this time I had worked as a volunteer, a necessary step in getting the movement off the ground, but one which had caused considerable financial hardship. Now that Clean Up Australia (CUA) had financial backing from its corporate sponsors, and Clean Up the World (CUW) had UNEP's support, it was time to change the structure of the organisation and draw a salary. Back in Sydney I handed over Alfred Builders to a young friend, Doug Rikard-Bell, and moved into our small Pyrmont office as full-time executive chairman. My loyal and hardworking secretary of many years in the building business, Jill Wunderlich, came with me, but to make it work — even on a shoestring — we needed an annual budget of over $1 million. To make it all work as it should we needed an annual budget of $3 million.

Our principal sponsor, Westpac, had continued to back us and their support grew in line with the expansion of the campaign. Our subsidiary sponsors — Amway, Comet and McDonalds — were equally enthusiastic and committed, each signing three-year sponsorship deals and providing some future security for the campaign.

Through our association with IBM, who supplied computers for both CUA and CUW, I was invited to Seoul, where South Korea's largest newspaper, *Chosun Ilbo*, was interested in sponsoring a cleanup. There I found immense enthusiasm from people at all levels, and the first Clean Up Korea Day was a huge success. By 1993, when Korea joined in Clean Up the World, almost one in four of the population — or nine million people — turned out to help.

Unfortunately, reactions elsewhere were not always positive, particularly from the richer, developed countries. In the United States, the Center For Marine Conservation — the same group whose support of Mark Schrader in the 1986–7 BOC Challenge had got me thinking about cleaning up Sydney Harbour — were convinced we had hijacked their ideas and gave us short shrift. The Keep America Beautiful council took the view that they were already doing what we did, but of course they weren't. Keep America Beautiful, funded by the packaging industry, is certainly a worthwhile outfit, but it had never taken the broad position on environmental problems that we took.

Despite these setbacks, by 1993 it had become clear that Clean Up the World was going to be an unqualified success, a worldwide environmental initiative instigated in Australia and directed, through registered organising committees, from our Sydney headquarters. UNEP was sufficiently impressed by the work we had done to award me with their top environmental prize, the Global 500 Laureate, in Beijing at UNEP's official World Environment Day celebrations on 5 June.

By September we had eighty participating countries and more than thirty million participating people, ranging from Korea's massive effort to the efforts of small but totally dedicated committees in such places as the Congo, where twenty volunteers collected a tonne of kitchen waste, and Senegal, where 200 Boy Scouts and Girl Guides took part. Although it strangely attracted little interest at home in Australia, Clean Up the World was big news in many countries, particularly in the developing world, and UNEP was delighted with the results. We had done it! And CUW was set to become an annual event.

CHAPTER SEVENTEEN

AUSTRALIAN OF THE YEAR

By late 1993 Clean Up Australia had developed into a large and complex organisation. At its core was this annual event, the first Sunday each March, when Australians from all walks of life would don gumboots and gloves and get out and clean up their own patch, taking responsibility for their own environment. This now involved some 700 organising committees in cities and towns around the country, all operating under the centralised control of Clean Up Australia headquarters in Pyrmont, Sydney.

In addition to this, Clean Up Australia had taken on the responsibility of managing and promoting Clean Up the World on an annual basis. We had begun to publish an annual national Rubbish Report which used statistics gathered on Clean Up Australia Day to paint a picture of Australian waste management practices, and we had taken on a whole range of community environmental advice and education projects which underscored our basic philosophy — that Clean Up Australia Day was not a quick fix, but rather a first step, a campaign drawing attention to the plight of the environment and acting as a catalyst for further action and long-term change.

In short, we had responded to community needs and provided a strong and independent voice on environmental issues, which opinion polls showed were more important to Australians than any other issue except employment. But Clean Up Australia went way beyond mere talk about these issues — we were in the business of activating people. And we were also getting into the business of solving environmental problems at their source, rather than cleaning up after them.

Australian Business Magazine's John Arbouw best summed up this subtle shift. 'Kiernan's new emphasis,' he wrote, 'moves beyond organising a reactive response to Australia's littering habits to a politically proactive stance which demands fundamental changes in how that rubbish is generated.'

I would be the first person to admit that my learning curve on environmental issues has been steeper than most. If, as my old school mate Graham Hartigan contends, I was an environmental vandal at seventeen, I wasn't much better at twenty-five, when we still regarded Sydney Harbour as a dumping ground for industrial waste. We used to throw corrugated iron roofs into it! An old builder had told me that if you threw roofing into salt water it would corrode away to nothing in a few weeks, and I was so uncaring about these things that I chose to believe him. When I was at sea I regarded it as my God-given right to throw my plastic bag of rubbish over the stern every four days. My attitudes probably started to change by the mid-70s, but it wasn't until the mid-80s that I did anything about them, and even into the 1990s, I was as ignorant as most about the major environmental issues confronting the world today.

The irony was that, as leader of Australia's biggest community-based environmental organisation, I was expected to know all about toxic incineration or the fine print of the Basel Convention. Furthermore, an Australian National Opinion Poll showed that Australians now trusted our organisation above all others. I didn't want to betray that trust by dispensing glib one-liners that masked an inadequate knowledge of the subject, so I set up an

Environment Advisory Committee to keep us informed on the latest environmental initiatives and breakthroughs, and to advise on the formulation of policy. Soon I was able to tap into the considerable vault of knowledge belonging to such people as broadcaster and scientist Robyn Williams, Professors Richard Bawden and David Russell, Tony Priestly of the CSIRO and Anne Prince of Recycle NSW. We would meet as a committee from time to time, but more often I would simply get on the phone and ask one of them to explain the Basel Convention or whatever it was.

I'm a builder by trade and a sailor by inclination, but I like to think I'm of no worse than average intelligence, and it doesn't take me too long to grasp a brief. Since we established the advisory committee, I think I've developed a basic, no-frills view on at least 100 environmental issues.

One of these which attracted my attention was sewage disposal and waste management. In late 1993 this led me to the waste water treatment system at Taronga Park Zoo. Like most people who had grown up around Sydney Harbour, I had (and still have) a deep and abiding love of the zoo, with its sandstone walls allowing animals to look out over one of the most magnificent views in the world, and ferry commuters staring up at the multi-levels of subtropical gardens and switchback paths. For me Taronga Park has always had an air of excitement and romance about it, but in 1993 I became aware that it also had more than its fair share of environmental problems.

Like much Australian industry, the zoo paid a licence fee to the Environmental Protection Authority and felt that this absolved it from any further environmental responsibility. The problem was that the EPA's standards were inadequate and the zoo's water treatment system and infrastructure were antiquated. They were within the law, but the law was allowing the zoo to dump animal effluent into the Harbour. When Mosman Council started monitoring pollution levels at all their harbour beaches, they discovered that Sirius Cove, below the zoo, was unfit for swimming. The band-aid measure they adopted was to remove the shark net to discourage swimmers, but when I heard about the problem I figured we could do better than that.

The way I saw it, there was no point in taking a revered institution like Taronga Park Zoo and beating it over the head in the media. A better approach would be to get Mosman Council, who were concerned about the excrement in the water, to sit down with the zoo and the EPA to see if we could work out a solution. It was an interesting experiment in environmental conflict resolution, and I was delighted to find that most parties were interested in finding an alternative system. Of course, not everyone saw a little bit of elephant shit in the water as any big deal. 'Throw some chlorine at it and extend the pipe a bit further,' said the young engineer from the Water Board. But we formed a steering committee and anyone who was at all negative found themselves hurriedly inducted so that they could learn something from people in industry who actually know how to solve problems.

After our initial discussions, I said: 'Why don't we fix this waste water problem and at the same time create an example of Australian waste water technology and storm water management that will show others in industry that developing cleaner systems can also save you money?'

Since we started Clean Up in 1989 I had developed a solid relationship with Memtec, an Australian waste technology company. I brought them in and also involved ANI-Kruger, another leader in waste water technology. We put together a $1.6 million waste water conversion system which, according to Taronga Park's executive director, Glenn Smith, is a world first for a zoo. Now, instead of running off filthy water into the Harbour, the zoo stores its animal excrement for use as fertiliser, while the waste water is recycled and used for watering gardens and cleaning out cages, thereby reducing the zoo's annual Water Board bill by forty to fifty per cent.

The Taronga Park project was an outstanding victory for the commonsense approach to solving environmental dilemmas, and real proof that Clean Up Australia could play a vital role in building bridges between industry, government and the general community. It also spurred us on to our next project, an ambitious undertaking we called 'Clean Up Australia 2001'.

With the sponsorship of Westpac again, we decided to ask our Clean Up committees and local government around the country to nominate environmental problems that needed solving, and we would see if together we could knock them off. Our aim is to identify 2001 problems like the waste water at Taronga Park Zoo, and either solve them or initiate programs to eliminate them by the year 2001. From the day we announced the project, our phones started ringing hot with nominations, but one of the first problems to emerge came not from an informant but from the devastating NSW bushfires of early 1994.

The fires revealed illegal dumps in many of our national parks. Concert promoter Michael Chugg had organised a bushfire benefit concert and generously passed on some of the funds to us so that, with the National Parks and Wildlife Service, we could organise a special cleanup. This was just another example of the community using Clean Up Australia as the agent of environmental action. Every day and in every way, I was beginning to see how wrong the politicians and bureaucrats had been when they sneered at our plan to clean up the Harbour. Individuals can make a difference.

With the help of the United Nations Environment Program, the ABC and Clean Up Australia in 1994 funded a television documentary which aimed to show how ordinary people were solving environmental problems around the world. I left Sydney with an ABC film crew and in just a couple of weeks we flew around the globe to six countries to make *Global Heroes.* In the Philippines we visited Smokey Mountain on the outskirts of Manila, a stinking massive hump made from a billion cubic metres of rubbish, where a Catholic priest was leading a movement which was beginning to turn a scavengers' slum into a profitable recycling and composting plant. In Germany we focused on consumers waging war on wasteful packaging which ended up in third world dumps like

Smokey Mountain. In Florida we looked at how one woman was fighting toxic waste which was destroying the river she had lived beside all her life, and in the Solomon Islands we met a courageous man who was establishing alternative employment schemes for logging-dependent villages.

The hour-long documentary that resulted was shown throughout Australia in the lead-up to Clean Up the World in 1994, and has since been shown around the world. It is a real document of hope, and I will be delighted if we can muster the resources to make more films like it, showing the resourcefulness, courage and determination of ordinary people who decide that they can change and improve things.

Late in 1993 I received a phone call from the writer and broadcaster Philip Adams, a wonderful bloke I'd known for some time. He said: 'What do these people have in common? Fred Hollows, Dick Smith, Patrick White . . .' He ran through about half a dozen famous names.

I replied: 'Well, they're all great Australians.'

'You're a great Australian too, Ian. You're Australian of the Year, in fact.'

When the shock wore off I probably said thanks, or something else that was totally inappropriate. Philip, in his capacity as chairman of the Australia Day Council, swore me to secrecy until the official ceremony, and I did pretty well, only letting the cat out of the bag to about six friends and family.

Australia Day dawned fine and clear, and apart from the slightly red ring around the sun from the previous week's bushfires, it was one of those picture-perfect summer days when you know why you live in Sydney. Judy could barely contain her excitement. 'Oh, Bicky, this is just fantastic,' she said at least a dozen times as we drove in a cab to Admiralty House for the official breakfast. And I have to admit, it felt pretty good. It wasn't quite as good as surfing down the face of a giant Southern Ocean comber under full sail, but it was pretty good.

As well as Judy, I took along Mum, Sal and Pip, Jack Earl and Kim and Giselle, the latter two because they deserved the award as much as I did. Jack had been one of the people I'd confided in when I was told I was to be named Australian of the Year, and to be truthful, the reason I let him in on it was that I wasn't sure Jack would last until Australia Day. His health had been up and down, but on the big day he looked twenty years younger than his eighty-five years, moustache waxed, hair slicked down and a gleam in his eye, as he too accepted an award, an Order of Australia Medal for his services to yachting and marine art.

Before I accepted my award from Prime Minister Paul Keating, Philip Adams introduced me as 'the Robin Hood of refuse, the Ghandi of garbage, the only garbo in the world more famous than Greta . . .'

After the presentation we were taken by limousine across town to Randwick racecourse where I was to sash the winner of the Australia Day Handicap. We sped through the city with a police motorcycle escort, the first time I'd enjoyed that privilege since I had a spot of bother with the constabulary one night in Honolulu!

Next stop was Tumbalong Park at Darling Harbour for the public presentation of the Australia Day Awards, at which the guest of honour was His Royal Highness the Prince of Wales. I've been an avowed republican for years. I think it's absurd that we should have a head of state 20,000 kilometres away, but at the same time I would like to see Australia retain its links with Britain, and I have some respect for Prince Charles as a person who cares about the planet. So I suppose I had mixed feelings about the event as I took my seat on the stage in the row behind Charles, his minder, NSW Premier John Fahey and other dignitaries. After I'd received my award I took my seat again and waited for Prince Charles to present some awards to school children. It was a balmy evening and I remember noticing for the first time just how big the crowd was. There were more than 20,000 people crammed into Tumbalong Park, casually dressed and in holiday mood, in something of a contrast to the guest of honour, who looked a bit stiff and uncomfortable in a double-breasted suit.

Suddenly I saw a figure break from the crowd and run towards the stage. My first thought was that this was some bizarre part of the festivities and at any second all would be revealed. Then there was the sound of a gunshot and a small puff of white smoke. The figure was still running at the stage and now I could see that he was a young Asian man who had a look of anguish on his face. He made a leap at the stage, caught his foot on something and fell, spread-eagled at the feet of John Fahey. Being the owner of a cattle dog, I am accustomed to instantly leaping into action to prevent disasters, and those instincts overtook me now. I sprinted around the line of chairs and grabbed the man in a headlock as he tried to get up. I could see the gun on the ground a couple of metres away, but it suddenly occurred to me that he might also have a knife, so I cranked up the headlock a notch or two.

At precisely the same moment that I leapt to my feet, John Fahey, a former rugby league player, had acted on instinct too, and tackled the assailant around the waist. He later told a reporter: 'I can recall focusing on a gun that was in the hand of some person who should not have been on the stage, and I wanted to get the gun. I grabbed the wrist and I held that hand until someone else took over.' Together now, the Premier and I lay on the stage, panting as the adrenalin began to flow, and wondering what to do next.

> *Kim McKay and I were sitting way up the back of the crowd. There was some confusion and some gasping from the front, but we had no idea what was going on. We figured that something bad had happened. I remember Kim saying, 'I guess this knocks Ian off the front page.'*
>
> Judy Kiernan

The prince's personal bodyguard, Superintendent Colin Trimming, and a small army of Special Branch police were soon swarming the stage. As John Fahey and I tried to extricate ourselves from the tangle of arms and legs with a modicum of dignity, a policeman stepped in and grabbed the young man in the most vicious hold I had ever seen, on or off the football field. It made my neck-cracking headlock look quite tame. As I got back to my feet I took stock of the situation for the first time. Flashes were popping from a hundred news cameras, people on the stage were in various degrees of shock, and, most importantly, the Prince of Wales was standing, quite unharmed, in the middle of a ring of protective flesh.

As the world now knows, the events of that Australia Day evening did not represent a serious assassination attempt — the assailant, David Kang, was armed with only a starter's pistol — but they were shocking enough in a society where we pride ourselves on public safety. When I later learnt the facts surrounding the attack, however, I had considerable sympathy for young Kang. Australian-born but of Korean extraction, he was a man who was deeply concerned about the welfare of the Cambodian boat people, and who could not get a hearing from anyone in a position of authority. Like a lot of people, he felt disenfranchised. Like the tragic, fanatical few, he had resorted to violence — or at least a caricature of it — to draw attention to his cause.

Prince Charles, who I believe had been blindsided and was still not fully aware of what had happened, continued with his children's presentation, and later did a walking tour through the crowd before adjourning to a private reception. I grabbed a beer and walked over to the prince, who was tucking into a gin and tonic in the company of John Fahey and Sydney's Lord Mayor Frank Sartor. I said: 'Congratulations, Your Royal Highness, on your remarkably calm approach to the incident.'

I know it's not protocol to report conversations with royalty, but bugger it! I've never had much time for protocol, and I reckon the prince owes me one anyhow. He replied: 'Well, you know, a couple of hundred years of breeding must be of some benefit.' He shot up in my estimation as he cracked a wry half-smile in appreciation of his own wit. Then he went on, 'You know, years ago in Africa I was on safari and a rogue bull elephant came out of the bush charging straight at me. I didn't have a clue what to do, so I stood there and did nothing, totally nonplussed. Do you know as it turned out this was exactly the right thing to do. The elephant stopped in its tracks, looked at me, put its ears down, its trunk down and walked off into the bush again.' Charles smiled again, a little more quizzically this time, and was led away to another group.

We finally made our escape and headed off to the Mosman Rowers', where Kim and my secretary Jill had organised a huge party in my honour. It had been a taxing, chaotic, emotionally charged day, and I think I really began to understand what this honour meant as we drove towards Mosman. More than ever before, I was public property now, a figure whose every move would be scrutinised and whose behaviour would be analysed as an example — good or bad — to others. Earlier in the day, outgoing Australian of the Year Mandawuy Yunupingu had been asked by a radio interviewer if he had any advice to pass on to me. He said: 'Don't get caught.' I think I knew what he meant.

The trick about it all was to keep my sense of humour — my mates would see to that — and to remember that I was nominated Australian of the Year on behalf of the many thousands of Australians who had made Clean Up such an enormous and important success. I was just the guy who got to wear the badge.

At the Mosman Rowers', perched above Mosman Bay where so many of my life's adventures had begun and ended, the manager, Henry Paterson, an old sailing mate, had organised a guard of honour with oars joined overhead, and as I walked into the club I was greeted with a huge cheer. We were more than an hour late and Kim had raced ahead and announced that this was because I had just saved Prince Charles. One could have been forgiven for assuming that this was a meeting of fervent monarchists, but I think the real reason for the raucous reception was that my mates had made a manful attempt to drink the bar dry before I arrived.

They were all there . . . both my wives, my precious and beautiful daughters, Mum, sister Ann, Jack, the Rat, Benny

Hawke, Zoran, Scaysbrook, the Crash and Burners, Rod and Kathy Muir, Kim and all the team from Clean Up, the Alfred Builders mob — the list was as long as the party itself. I looked around the room that night, and out the windows at my beautiful *Maris* sitting in the bay, and reflected on how lucky I'd been in life so far. I felt satisfied, but determined to push ahead with the next challenge. And I envied no man.

CHAPTER EIGHTEEN

A THOUSAND YEARS

My spirit is with each one of you right now for a toast. A toast to you for adding such tremendous beauty and interest to the great life I have enjoyed so much. I am more than proud of the great effort you are putting into your lives and for gaining so much from knowing you through the years.

After this drink I will be off and have no sadness in saying see you all later. Have another Scotch. All my love.

Jack Earl, 1994

No one can last forever. He was my dearest friend and in recent years we'd spoken about death. Once he said, 'Matey, I'm ready to go.' Not willing, perhaps, but ready.

In the years since Kath had gone 'into another dimension', as Jack liked to put it, he had busied himself with his painting and there had been great moments, like the return of *Kathleen Gillett* and, I suppose, the thrill of seeing me, his spirit son, named Australian of the Year. But how do you replace a soul mate like Kath? Jack had the love of his daughter Maris and the rest of his family, plus a huge network of friends who thought the world of him, but the passing of the one great love of his life caused a hurt so great that only a man of Jack's indomitable spirit could have carried on. And by the early months of 1994, he was finding it difficult. There was some physical deterioration, just a barely perceptible creeping tiredness that was slowly overtaking him.

I had taken to dropping into the studio at the Mosman Bay compound more frequently. We'd sit there, surrounded by his work, a bottle of whisky on the table, and we'd talk into the night. Jack never became trivial or distracted with age: in fact his wisdom seemed to grow. I'd bounce ideas off him as we drank, and I'd come away invigorated and inspired.

One Friday night in early April I called in for a drink and stayed very late. Jack seemed reluctant to stop the conversation, and when I did finally drag myself away, he fixed me with a steady, loving and serene look that sent chills through me, because I knew it so well. It was the expression I had seen on Kath's face when I sailed away towards Rio in 1987, never to see her again.

I had to travel interstate the following week and Jack had a fall on the Monday night, breaking his hip. He went into hospital for an operation, which was successful. The family visited him in post-op and he was weak, but still seemed okay. As he lay in his hospital bed, the launch of the fiftieth Sydney to Hobart Race took place on the dock at the Australian National Maritime Museum, next to *Kathleen Gillett.*

I flew home on the Wednesday morning and as soon as I got into my car and switched on my mobile phone, Judy was ringing. 'I'm coming out to the airport to pick you up,' she said. I knew immediately that something was wrong. I told her I was already on the road and that I would meet her at the office. When Judy walked in she didn't have to say it. I knew Jack was dead.

> *He just stopped breathing and that was that. Maris was called back to the hospital in the early morning and the Samoan nurse told her that she had tried to revive him with mouth-to-mouth*

but failed. In our time of sadness we had a quiet chuckle over this — the image of Jack fading out of this life being kissed by an islander girl. We imagined he would rather have liked that.
Ben Hawke

Jack had left a large wad of cash and some very specific instructions about how his funeral and wake were to be conducted. He wanted to be placed in a plain pine box without flowers but with kelp instead, so I went out in *Maris* with his grandsons Benny and Matt, and we dived for kelp off Bradleys Head, making black jokes about killer sharks and blue-ringed octopuses, in the compound's practical approach to dealing with grief.

The funeral service at the church was short and succinct, but the wake back at the compound was a wild, tearful and yet joyous affair. True to Jack's wishes, I had engaged the Ocean Paradise Quartet, a bunch of melodic Cook Islanders, and we drank our whiskies and sang along to the old songs of the Pacific.

Mick Earl came down from Canada a few days later and I took the children and the grandchildren out on *Maris* to see the old boy off. Mick carried the ashes and I brought along seven hibiscus flowers from Mosman Bay, one for each of us on the boat that flat, windless autumn morning. As we sailed out of the bay we were joined by some fairy penguins, then eight dolphins appeared and seemed to escort us out of the Heads. We went out six nautical miles and Mick let the ashes go. Then we poured a bottle of whisky over the water and threw our flowers in. As the tears flowed, I noticed our sailing mate Dick Hammond standing off in his yacht about 500 metres away. It was a fitting tribute from a mate, not wanting to intrude on the family, but wanting to pay his final respects.

One of the grandchildren said, 'I'm in for a swim!' In a minute we'd all stripped off and jumped in, laughing, crying and shouting for joy all at once. Maris's husband Tommo, a big, gentle Pole, swam over to me and said with a laugh: 'This is a strange ritual.' And I suppose it was, but somehow apt.

When we left to come back in I gave the tiller to Mick and a breeze sprang up from nowhere. I popped a little spinnaker up and we shot back over the floating flowers and into the Harbour and home. It was a beautiful sail, and it was over.

Sadness always travels in company, and in July we learnt that Peter Livingstone Curran, the Bilge Rat, had serious cancer. I went alone and sat at the chart table of *Maris* down in Mosman Bay, right where we'd fitted out *Bimblegumbie* all those years ago, saved her from sinking and enjoyed a thousand beers and 10,000 yarns, a stone's throw from where we'd done the deal to buy *Maris*, the Rat's old cave home in the cliff and Jack and Kath's garden overlooking the bay. I sat there and drank in sorrow while I leafed through the bits and pieces I'd jotted down about the Rat over the years.

In my Blue Book I found a sheet of 'Raticisms', of which my favourite was the classic, 'I think you've got your ambitions mixed up with your capabilities.' I also found a brief article I'd written about him for the yachting magazine, *Sydney Afloat*, a few years earlier:

> One Monday I was slipping my lines and clearing for Lord Howe when I passed the Rat, sitting on the sea wall drinking coffee.
>
> 'Where ya going?'
>
> 'Lord Howe, wanna come?'
>
> 'I'll just get my gear.'
>
> Five minutes later he walked off the end of the wharf with a bag of T-shirts onto the *Maris*. After two weeks we sailed back into Mosman Bay with adventures that would fill a small phone book . . . [The Rat] was a gypsy with no respect for authority or the formal system; but he was clever and wise. He was also very thirsty, since, it is claimed, he swallowed a Wettex as a child . . . He says one day there will be two sets of foot prints down the beach towards where the *Maris* was.

I pulled another rum from the old keg and wrote my friend a letter:

> Dear Rat,
>
> I want to tell you what I really think of you! Since the days of *Bimblegumbie* when our friendship fired into life, the humour and irreverence that we have shared has been an absolute delight. The things we have shared — the Pig, the Porsche, the Siamese Honda 4 and, of course, the *Maris*... Thanks for all your help with her... I told you the other night I have always regarded you as a big man. Your strength and courage in this fight make you a giant in my eyes. I love you as my brother, and you are that. Two sets of footprints in the sand. We are with you, mate, with love and luck forever.
>
> Bicky

The Rat put up a gallant fight and never lost his spirit — death was never an option for him — but in the end it was almost unbearable to watch his life slowly ebbing from him while he suffered such pain. I am not a religious man, but several times I sat by his hospital bed and prayed for his release from this torture. One night he even begged me for a gun to end it all. Mercifully, in early March 1995, the Rat slipped his lines and cast off for a better place.

Henry Patterson from the Mosman Rowers' and his wife Penny are great and generous people, and they had taken the Rat into their home in his final year, easing him through those painful months. Now they drove up to the north coast to bring Pete's parents back for the funeral. So many of the people whose lives the Rat had touched stood around and waited for the casket to arrive. Henry whispered in my ear: 'He's bloody late for his own funeral!'

In due course the Rat arrived and I gave a simple eulogy: 'We are here to say farewell to one of our own, Peter L. Curran. A blue-collar philosopher, a tradesman of great skill. The mould is now broken.' I quoted a few 'Raticisms', like 'you can't ride the top spoke forever', and 'the gaze of the eye is much stronger than the shake of the hand'. I also mentioned that he was not averse to plagiarism, and had often borrowed my line, 'I am not here to buy a suit.' The congregation got a chuckle out of that.

Then it was farewell, old mate, and back to the Rowers' for a wake and many Rat-tales.

We intended to scatter the Rat's ashes in the ocean and a growing mob gathered on the *Maris* one Sunday morning. As fate would have it the motor on the *Maris* failed to fire and the decision was made to scatter them there and then. His ashes now rest under the keel of the *Maris* where she lies in Mosman Bay. Appropriate, I feel.

The loss of two of my closest mates inside a year knocked me around somewhat, but it also had the effect of making me stop running long enough to consider what sort of a person I'd become in my middle years, what sort of a life I'd had thus far and what lay in the future. More importantly, it has made me look closely at where I want to go, personally and professionally, over the next ten years as we enter the new millennium.

At Clean Up Australia I want to put in place a central foundation of funding which gives us the financial independence not only to clean up but fix up this country, to repair our environmental assets and to return them to a community charged with future responsibility for them. To do this we need to encourage industry to provide the methods to manage our wastes, both human and industrial, and to provide these technological solutions as a commercial service. It is crucial that we convince industry to become a serious player in the environmental arena, reaping commercial benefit from its contribution to 'fixing' the environment.

At the same time, I am totally opposed to the privatisation of public utilities. We need to harness the strengths of industry to be on our team, rather than have it acquire publicly-owned services as profit producers. We have to take a 'big picture' perspective and actually start calculating the value of the environment as an asset that belongs to all of us.

I want to take the Clean Up organisation to a new stage in which we communicate through education, making documentaries which spread the message of repair globally. Moreover, I want to make our organisation truly global so that as one people we can seek solutions to our environmental problems. And I want to continue my sailing career with an environmental flagship that carries out research projects and acts as a travelling classroom for all ages.

I have truly enjoyed my voyage of adventure through life so far. To me there has always been a sense of urgency about it, and I have tried to cram so much into every moment. Whether it be trading in a rising property market or running before a gale in a rising sea, I have always liked to push my boat (or myself) to new levels of performance. But it has made things difficult for those around me when the wind dies to nothing, or the market crashes.

I am not a fair weather sailor. I like to push the edge of the envelope, and if that means in my life I have had to suffer the consequences, then so be it. I know no other way. And that attitude has enabled me to build a fortune (and lose it) out of derelict buildings and help change the social and architectural face of my home town. It has enabled me to guide a stubborn boat through the oily calm of the doldrums and sail on through treacherous seas around the world. And it has enabled me to pick up a garbage bag and say to the world: 'One person can make a difference.'

The people who have made a difference in my life are many and varied, but of course some spring immediately to mind. My dog, my boat and the oceans of the world are not really people, but I love them none the less. I love my wife, Judy, my wonderful kids who are such valued friends, and I even still love my first wife for her determination, her tenacity and her efforts in raising the girls when I was off hurling myself at life's adventures.

I love and value my few close friends, and the loss of them is something I find very difficult to deal with. I think of Jack every day of my life, and the loss of him and the Rat has left a hollow spot.

But I guess you go to more and more funerals as you approach your own.

EPILOGUE

ROARING DOWN TO HOBART

While *Maris* had done a couple of Sydney to Hobart races under Jack Earl, I had never skippered my own boat in one. When Jack passed away just as the fiftieth anniversary running of the race he helped found was announced, the vague thoughts I'd had about starting *Maris* in the special veterans section of the race suddenly became a firm personal commitment. Despite the huge demands on my time during my reign as Australian of the Year, I would do it for Jack.

Putting a crew together was the easiest part. I phoned Tommy Stephenson in Brisbane and asked him to be sailing master. He said: 'How quickly?'. Jack's grandson, Matty Tomachevsky had developed into a fine young sailor and he was in, as was my old mate Dick 'Sighty' Hammond, a veteran of thirty-five Sydney-Hobarts, and the finest tactical navigator I know. Then I phoned Marty Farrand in Auckland. Yes, he would come over with the family and make a summer holiday of it. Now I had a team of good mates and fine sailors, but there was one slightly complicating factor.

Doug Rickard-Bell, who had taken over Alfred Builders from

me, was a sailing rookie but a fine fellow who was eager to learn and game for anything. He'd sailed to Lord Howe Island with me and asked now if he could be first reserve. I said of course he could, but as the race drew nearer his enthusiasm was such that I couldn't let him down. I phoned Tommy and said: 'What about we take six?' Tommy said sure.

Maris was more than thirty-five years old now, and needed a lot of work if she was going to be competitive, so I sought sponsorship. I suppose if I couldn't get sponsorship in 1994, I was never going to get it, and sure enough the Canon corporation, who'd been very supportive of Clean Up Australia, came to the party. Their chief executive in Australia, Mr Miyagi, was delighted by my proposal to rename the boat *Canon Maris* and to carry the Canon colours on her mainsail.

Although there was a dreadful panic to get her finished in time, *Canon Maris* motored to the starting line on Boxing Day looking better than she had in all her years — new rigging, sails, deck, electronics and winches. Standing in the cockpit, we were fired up like excited kids, even though we had more than fifty Sydney-Hobarts between us. (Don Mickleborough's crew on *Southerly* had 110.)

It was a spinnaker start under grey and overcast conditions, a moderate sou'easter and a lot of chop on the water, but we started well, third across the line. As we made for the first mark the snap-shackle on the spinnaker let go, and we had to take it straight off. There was no point trying to set it again just a couple of hundred metres from the mark. Just as the spinnaker came down, *Ondine III*, a huge maxi, came bearing down on us, stealing our wind and making the going very difficult. We muttered a few curses, but in a way it was poetic justice. *Ondine* had given me my start in Sydney-Hobart races, crewing with the Americans back in the early '70s.

We cleared the Heads and settled the boat down. Eventually the breeze backed around and the next day we had our spinnaker on and started to roar down the coast to Bass Strait. *Maris* was never going to surf like a modern boat, but by God the old girl excelled herself. We were clocking lots of ten-knot waves and I was particularly chuffed when I surfed her down one at 11.7 knots. Then, with Tommy at the helm and Matty up on the foredeck thinking about hanging five, we surfed down the face of a monster, bow wave out front, rooster tail out the back, at 13.7 knots, a new record for *Maris.* We hooted and hollered and danced about like kids.

We started to move up through the fleet and stood in seventeenth position on corrected time as we entered Bass Strait. A southerly front was on its way and on deck I noticed the cloud gathering on the horizon. It was moving towards us very quickly. I called Tommy and Dick up on deck and asked them for their opinions. We had been talking about how boats in this race get caught with their spinnakers up and have to run back towards Sydney to get them off. It had happened to the maxis one year and we didn't want it happening to us. Marty was checking the progress of the front with his forefinger. He said with a note of urgency in his voice: 'It's coming very fast.'

It was time to move. We got the spinnaker off, set the No 4 jib, took the main off altogether and just sat with jib and mizzen. Then it hit us — a roll of cloud with whiskers off its top, roaring across the ocean in this strange yellow light. As it hit, the wind went from 35 to 40 to 50 to 60 knots, turning the sea into a smoking field of white caps. But we were snugged down well on the wind, and we made good progress while other boats did not.

Off the Freycinet Peninsula the breeze dropped away to nothing. In this race in which it is possible — even probable — that you will experience the full gamut of weather conditions over a few days, this was exactly what we had done. For more than eight hours we suffered the light conditions that *Maris* loathes. Sometimes we had 'lemons' (three noughts) on the speed dial as we tried to urge performance out of the boat.

Somehow we made it to Tasman Island, where we found that *Cherana* and *Southerly* were the only veterans ahead of us. This still gave us a chance on handicap, but we were really after line honours. As we entered Storm Bay we had to make the decision that could win or lose the race. We bowed to Dick's experience and plotted a course down the port side of the bay, where we picked up better breeze and current. Now we were well placed and the excitement mounted again as we made for the Derwent.

Inside the river we were met with forty-five knots of breeze again. It blew like hell but we sat the boat up and, with lots of headsail changes, crossed the finish line at 5.20 pm on New Year's Eve, the last day of the biggest year of my life.

We were third across the line in five days and four hours, three hours behind *Southerly* and half an hour behind *Cherana*. On corrected time we were second to *Southerly* by just seven minutes. Overall we finished sixtieth in a fleet of 370.

We had a traditional quiet little drink in Hobart on New Year's Day, with Jack looking down on us, I'm sure. We toasted each other's health and raised our glasses to *Maris* for a gutsy performance from the old girl. Then Marty Farrand threw a pie in my face, which everyone thought tremendously funny.

I'll get the bastard back. There's life in the old dog yet.